Sorourozaman Shakeri Torshizi

Review of the Application of Nanotechnology

Sorourozaman Shakeri Torshizi

Review of the Application of Nanotechnology

in the Food Industry

Scholars' Press

Imprint
Any brand names and product names mentioned in this book are subject to trademark, brand or patent protection and are trademarks or registered trademarks of their respective holders. The use of brand names, product names, common names, trade names, product descriptions etc. even without a particular marking in this work is in no way to be construed to mean that such names may be regarded as unrestricted in respect of trademark and brand protection legislation and could thus be used by anyone.

Cover image: www.ingimage.com

Publisher:
Scholars' Press
is a trademark of
Dodo Books Indian Ocean Ltd. and OmniScriptum S.R.L publishing group

120 High Road, East Finchley, London, N2 9ED, United Kingdom
Str. Armeneasca 28/1, office 1, Chisinau MD-2012, Republic of Moldova, Europe
Managing Directors: Ieva Konstantinova, Victoria Ursu
info@omniscriptum.com

Printed at: see last page
ISBN: 978-620-8-84712-8

Review of the Application of Nanotechnology in the Food Industry

By

Sorourozaman Shakeri Torshizi[1, 2]

[1]*Department of Food Science and Technology, Faculty of Agriculture, Ferdowsi University of Mashhad. Mashhad, Iran*

[2]*Department of Biochemistry, Faculty of Biology, Payam-e Noor University of Mashhad, Mashhad, Iran*

Content

Chapter I

Introduction

Introduction to the subject

Nanotechnology is the ability to produce new materials, tools and systems by taking control at the molecular and atomic levels and using their properties at the nanoscale. Nanoscience is the study and research of devices and structures that exist in the smallest unit of dimension (200) nanometers or smaller. From the above definitions, it can be seen that nanotechnology is not a discipline but a new approach in all disciplines. Applications for nanotechnology have been listed in various fields from food, medicine, medical diagnostics and biotechnology to electronics, computers, communications, transportation, energy, environment, aerospace materials and national security: The wide applications of this field and its social, political and legal consequences have raised this technology as a multidisciplinary and multidisciplinary field.

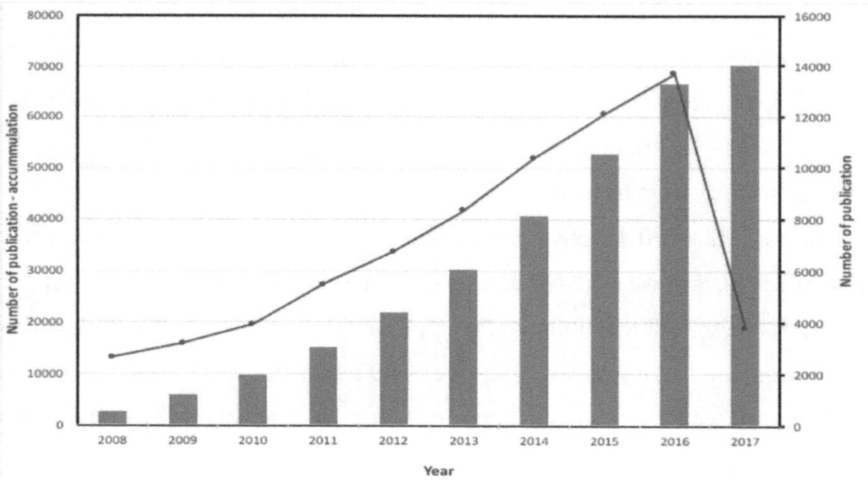

Figure 1. Nanotechnology in Sustainable Agriculture

Although experiments and research on nanotechnology have been pursued seriously since the beginning of the 20th century, the transformative, miraculous and incredible effects of nanotechnology in the research and development process have attracted the attention of all major countries and have considered nanotechnology as one of their most important research priorities during the first decade of the 21st century. Japan spent $400 million in 2001 and $960 million in 2004, and the United States allocated about $3.7 billion for this purpose in 2005-2008.

The use of this technology in all sciences has made nanotechnology research a major scientific and industrial challenge facing the world. Therefore, Iranian researchers, professors, and industrialists must also identify their position and position on this issue in a public mobilization and create an active and even competitive presence in this position. For such a task, designing a comprehensive and comprehensive program is inevitable. Nanotechnology has three branches: Dry, wet, and computational nanotechnology, which are used in various sciences, especially in the manufacture and production of electronic, medical, and food industries.

History of Nano in the World

Forty years ago, the theoretical quantum expert and Nobel laureate Richard Feynman, in his famous 1959 lecture "There's a Lot of Space down there," explored the undeveloped dimension of materials science. He stated at the time that the principles of physics, as far as I can understand them, do not say anything against the possibility of building things atom by atom.

He hypothesized that if scientists learned how to build transistors and other structures on small scales, then we would be able to make them smaller and smaller. In fact, they would be close to their true limits on the unknown quantum edge, so that if we put one atom against another in such a way that we could create the smallest possible artificial and fabricated product, the question remains: What devices could we make using these very small forms?

Feynman envisioned a molecular doctor hundreds of times smaller than a single cell that could be injected into the body and used to perform tasks, study and verify the health of cells, perform repairs, and generally maintain the body in perfect health. The term nanotechnology was first coined by Noriota Yoneguchi, a professor of science at the University of Tokyo.

He used the term to describe the creation of precise materials (devices) with dimensional tolerances in the nanometer range. Minsky was able to give Feynman's ideas a boost. Minsky, the father of artificial intelligence, and his student Drexler brought together a group of computer science students in a society. He engaged the minds of the younger generation with a set of ideas that he called nanotechnology. Drexler received his only Ph.D. in nanotechnology in 1991 from MIT. He is a pioneer in the nanotechnology project. Many of the major technological advances of the 20th century, including information technology and biotechnology, would have been disrupted without the use of nanotechnology.

Basics of Nanotechnology

Nanotechnology is a fast and new technology that allows the work, manipulation and production of tools, materials and structures at the molecular and even atomic level by atom in functional structures in the nanometer dimension. With nano methods, we can create multifunctional, self-regulating, self-controlling and self-healing devices. The word nano is derived from a Greek word meaning "Small or tiny". Nano is a size between 1 and 100 nanometers. The wavelength of visible light is between 400-700 nanometers and a living cell is equivalent to one micron (1000 nanometers).

One of the most important sectors of the industry of all countries that is related to food security is the food industry. With the shortage of food resources and the increase in population, the development of this sector seems essential. The use of new technologies in this sector is a new approach that is receiving much attention. Our country is one of the largest importers of food products; therefore, if we ignore food production and processing, the amount of food imports will increase every day. Nanotechnology offers the food industry several new methods to improve the quality, shelf life, safety and health of food. However, there are concerns among consumers, regulatory organizations and the food industry about the potential adverse effects (toxicity) associated with the application of nanotechnology in foods.

In particular, a major concern is the direct incorporation of engineered nanoparticles into food, such as those used as flavorings, preservatives, nutrients, or those used to modify the optical, rheological, or flow properties of foods or in food packaging.

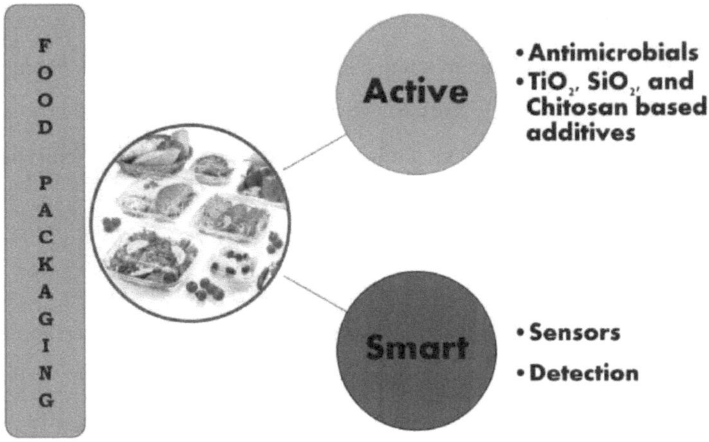

Figure 2. Application of nanotechnology in different aspects of the food industry

Food safety alongside consumer health

According to food industry experts, in the first perspective, packaging is considered a solution to increase the shelf life of food, and various methods can be used to achieve this goal.

Dr. Bahareh Emadzadeh, Director of the Food Nanotechnology Group at the Food Science and Technology Research Institute, said in an interview with Jam Jam: "One of the methods that has received attention in many countries and unfortunately remains at the research level in our country is the use of nanotechnology capabilities. We can design indicators based on this technology that monitor the shelf life of food products. The function of these indicators is that, for example, colored indicators are embedded on them that change color over time.

Now these indicators or indicators are produced without using nanotechnology. Studies and experiments show that the use of nanotechnology can significantly increase the sensitivity of these indicators. Nanoparticles have a specific structure, and this structure can increase the sensitivity of indicators in monitoring the shelf life of food." Apart from these, nanomaterials and products produced from nanomaterials, due to their unique properties, can play a role in controlling the inhibitory properties of packaging and control the gases entering food packaging.

In Iran, we have not been able to progress in food packaging in line with other sectors of technology. That is, good steps have been taken and significant progress has been made in many of the technologies in the field of food science and industry, but one of the sectors that has been neglected and has not received special attention from an industrial perspective is packaging and the technologies that can be used in this field. Although environmental safety and the use of environmentally friendly materials in food packaging are important, this trend must continue to the extent that consumers are not harmed. On the other hand, it must be ensured that the residues of nano compounds that enter the wastewater do not endanger human health and the environment. These are some of the issues that have become controversial in this field. Dr. Emadzadeh believes in this regard: Every science should be examined from two aspects; both in terms of its role in the health of society and in terms of the risks it may bring with it. In every science, the positive aspects first show themselves more, and over time and in studies and research, the negative aspects gradually become apparent. Packaging can be in the form of a coating that comes into direct contact with the product. Sometimes, packaging is not in direct contact with the product. There is a risky view regarding food products that use nanotechnology in their packaging, but the harmful effects of the presence of these compounds in nature on the environment is a topic that must be studied in depth.

Active Packaging

Our country has not made much progress in the field of packaging, and if anything has been done in this field, it has been limited to academic articles. One of the works that has recently been considered at the Food Science and Technology Research Institute is a special type of packaging that carries the attribute of active. Smart packaging is a more advanced type of active packaging.

Research on packaging with this feature is not very old, but despite this, significant progress has been made in this field, and various types of active packaging have now been commercialized in some countries, including the United States and Japan. According to Dr. Razieh Nezam, a member of the Food Chemistry Department of the Food Science and Technology Research Institute and the executive of the active packaging project, this research institute has been working on a special type of active packaging that is an oxygen absorber.

In this packaging, a bioactive material is used that is placed inside the packaging instead of directly contacting the food material or product and performs the function expected of it inside the packaging. Usually, the main purpose of these packages is to increase the shelf life of the food, and the important point is that this compound is included in the packaging and may be released over time and enter the food product. Therefore, this material must be safe from the food industry's point of view and not cause any problems.

This group of researchers has used ascorbate salt, which is a type of vitamin C, in active packaging. In addition, this salt is an antioxidant and oxygen receptor compound. Even if it is released and even a large amount of it enters the food, it does not cause a problem. Of course, in terms of technology, the work must progress to semi-industrial and industrialization, but the advantage of this method is that not much work has been done on it in Iran.

Many compounds in the nanoscale are dangerous because nano is defined based on different characteristics of materials in smaller dimensions. In response to the question of whether research has been conducted on the safety of ascorbate salt in the nanoscale, Dr. Nezam said: We have generally conducted research on this compound in the nanoscale. Using materials in the nanoscale increases the uniformity of the work. Many particles in this size are problematic, but ascorbate salt is a compound that is safe even in the nanoscale, and this compound may even be present in the food itself at such a scale. Some compounds used in the field of nanotechnology may not be safe.

In future work, we plan to use other antioxidants and extracts that have different medicinal properties. The introduction of these compounds into the packaging will make the packaging active and exhibit a specific function. This packaging can be used for a wide range of products.

A group of researchers believe that the safety of nanomaterials is not confirmed and can harm human and social health. Another group says that we should move in this direction to keep pace with other countries and remove existing obstacles. Predictions show that in the near future, a day will come when we will be able to prepare the food we need from atomic and molecular compounds. Although there is a long way to go for the practical use of this technology, it is expected that this progress can lead to a path to expanding the production of higher quality and more sustainable food products, which will not only allow us to use higher quality food, but will also save on raw material consumption.

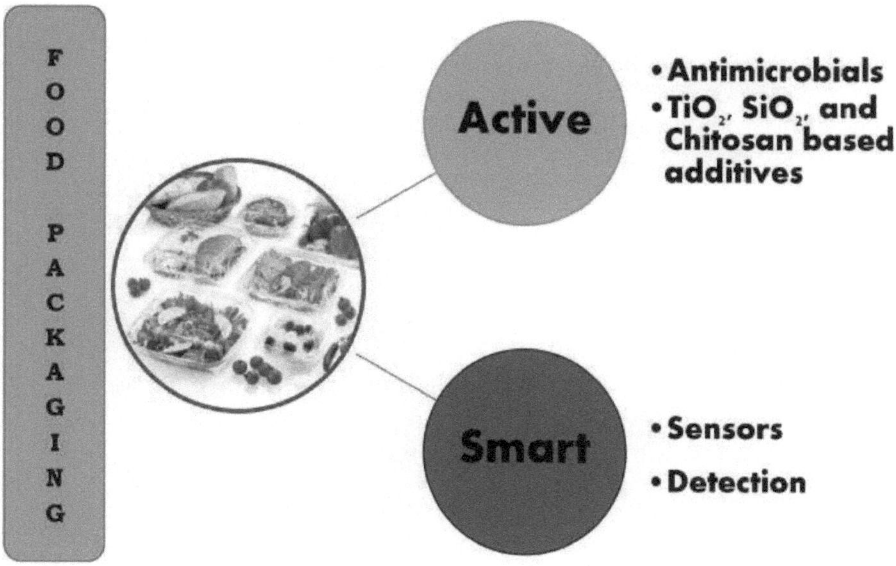

Figure 3. Nanotechnology-based approaches for food sensing and packaging applications

Chapter II

Review of the application of nanotechnology in the food industry

One of the most important sectors of the country's industry is the food industry, which is closely related to the food security of the people of the society. The increase in population, along with the expansion of urbanization and the increase in per capita income, has increased the need for processed foods day by day. Therefore, the use of new technologies, including nanotechnology, in this industry has attracted much attention from scientific and industrial circles around the world.

The various fields of application of nanotechnology in food and food industries can be divided into the following six categories:

- ✓ **Food preservation**

The use of nanotechnology in food preservation is one of the important solutions to increase shelf life and reduce food waste. This technology, using nanomaterials and nanoparticles, offers new capabilities for improving food preservation processes. By using nanoparticles, we can increase the shelf life of food because these particles can act as antimicrobial agents and reduce the growth of microorganisms. By improving the shelf life and reducing food contamination, this technology can help reduce food waste and be used to increase the productivity of food resources.

- ✓ **Improving flavor and color**

By using nanoparticles, natural colors can be produced with high quality and strength. These colors can be used to increase the visual appeal of food while avoiding the problems associated with the use of industrial colors that may be harmful to health. Also, through the use of nanoparticles, aromas and essential oils in food can be improved and controlled. This leads to an increase in the amount of desired aroma and flavor in food products.

- ✓ **Food Safety**

Nanotechnology in the food safety field has brought about significant changes in the production, safety and quality of food. This technology is widely used in various stages from food production to consumption. Food nanoparticles can effectively help protect antioxidants and vitamins, which increases the resistance of antioxidants to processes caused by heat, light and oxygen. Antibacterial nanoparticles can also help protect food from spoilage without the need for high amounts of preservatives.

- ✓ **Packaging**

Nanotechnology in food packaging is one of the important and innovative areas in the food industry that provides new possibilities for increasing the quality and safety of food by using precise measurement and control of the dimensions and properties of materials at the nanometer scale. This technology is used as a powerful tool in improving the characteristics of food packaging, including maintaining freshness, preventing spoilage, increasing shelf life and reducing waste.

The use of nanomaterials in food packaging can reduce environmental damage, as less packaging material can be used, resulting in reduced weight and volume. The technology can also help improve the performance and safety of packaging, for example by increasing resistance to air, moisture and food penetration. For pollution control, the use of nanomaterials in food filters and packaging can help reduce biological and chemical contamination and protect food from contaminants.

- ✓ **Food production**

The use of silver nanoparticles, titanium dioxide nanoparticles, etc. as antibacterial agents in food packaging can extend the shelf life of food and prevent the growth of harmful microorganisms due to their high antimicrobial activity. The size of the nanoparticles also allows nutrients to be absorbed more effectively in the body. For example, increasing the surface area of nanoparticles of nutrients such as vitamins and antioxidants can increase their absorption by the body. The application of nanomaterials in food production can improve their physical and sensory properties. For example, the use of nanoparticles in ice cream production can increase structural stability and reduce the formation of large crystals in the product.

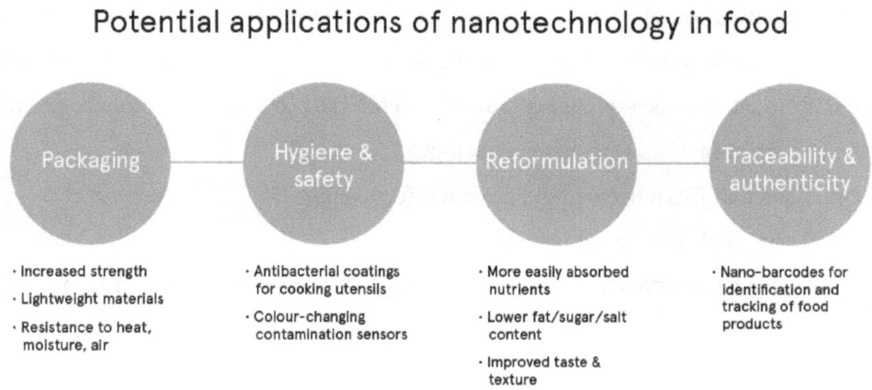

Figure 4. Opportunities for nanotechnology in food and feed

Application of nano powders in food processes

Nanotechnology in the field of food processing is an important and exciting development that brings significant improvements in the production, packaging, safety and quality of foods. The use of nanoparticles in colors and flavors can provide significant improvements in the appearance and flavor of foods. In addition, mineral nanomaterials such as titanium dioxide nanoparticles can act as safe additives to replace synthetic pigments in foods.

The use of nanotechnology in food packaging can also increase the shelf life, reduce contamination and maintain the quality of food products. Nanomaterials such as nano silica can be used as packaging strips and improve the storage and quality properties of foods. In rapid detection of contamination, nano sensors, carbon nanotubes and nanotechnology-based diagnostic devices can quickly and accurately detect food contaminants such as bacteria and viruses, which can help improve food safety. Adding nanomaterials such as nano calcium carbonate to food products can increase their nutritional appeal. For example, adding calcium carbonate nanoparticles to dairy products can increase nutritional appeal and reduce the need for adding artificial calcium. Now let's examine the role of nanotechnology in food production.

1. Creating foods with new flavors

To treat many diseases such as diabetes (sugar), a special diet must be used. On the other hand, due to today's industrial life, the issue of obesity has also become a problem that unfortunately affects adolescents and even children. To solve this problem, several solutions are proposed using nanotechnology:

- ✓ Food that fills a person but does not affect weight.
- ✓ Tasty foods that contain fat substitutes.
- ✓ Using nanoparticles to prevent the body from absorbing and storing fat and calories.

2. Producing enriched foods

Some materials have valuable components that are very important and essential for the body, but for reasons such as taste and habit, there is not much desire to use them. If we can separate these valuable substances and add them to other foods as additives, we can make fortified foods. Nano filters have been developed that screen molecules based on shape rather than size, allowing us to separate specific components of a product.

3. Molecular Food Production

Researchers believe that in the future, molecular engineering will make it possible to grow and produce large quantities of food without the need for soil, seeds, farms, or farmers, and that with the advent of this technology, world hunger will be eradicated. With this idea in mind, instead of growing grains and raising cows to obtain carbohydrates and protein, nano machines will create our desired steak or flour from the carbon, hydrogen, and oxygen atoms in water or carbon dioxide in the air. Nano bots in food will also move through the bloodstream, cleaning it of fatty residues and deadly pathogens.

Molecular production of food is one of the goals and aspirations of nanotechnology and it seems that it will not be achieved very quickly. By using molecular foods, hunger can be eradicated, the nutritional content of foods can be increased and the risk of allergens in foods can be eliminated.

4. Food additives at the nanoscale

Nanoscale food additives are substances that, due to their very small size, can have very effective effects on the nutritional properties as well as the sensory and textural characteristics of foods. These additives are often obtained from natural substances such as vitamins, minerals, plant extracts and various nutrients that are added to foods in nanometer form.

For example, BASF has produced a type of carotenoid in nanoscale. Carotenoids are a type of food additive that gives foods their orange color and are naturally found in carrots and tomatoes. Some types of carotenoids are antioxidants and are converted into vitamin A in the body. BASF sells this carotenoid in nanoscale to large food and beverage companies around the world to be used in lemonades, juices and margarines. Formulating additives at the nanoscale makes them easier to absorb in the body and increases their shelf life. BASF has also developed another industrial carotenoid called lycopene as a food additive. Lycopene occurs naturally in tomatoes.

5. Foods with specific release in the body

In order for our body to benefit from the release of food components, the nutrients must go to a specific location in the body and be activated there. Controlling and engineering the release of nutrients in the body is one of the research areas of nanotechnology. These foods, which are called "Food and medicine", have their active components distributed in the body by nano capsules. Because one way to preserve an active food component is to place it in a protective coating. This coating can be designed so that it dissolves when stimulated by a suitable stimulus and the active substance inside it is released through the coating.

For example, the George Weston Food Institute in Australia has produced a type of bread called Tip Top Up, which contains omega-3 fatty acid oil from tuna. However, the tuna oil is placed inside microcapsules, so the consumer does not taste the fish oil and is only released when the oil reaches the stomach and the capsule is digested. This technology has also been used in yogurt and baby food. By using nano capsules made of edible polymer, the taste and smell of food molecules can be preserved, thereby increasing the shelf life of the product.

6. Enzyme coating

One of the concerns of the world's food industry companies is the preservation of food and its protection from enzyme damage. If enzymes can be removed from the food environment in some way, the process of food spoilage will be delayed. Using nanotechnology, enzymes can be removed from the environment of activity and prevented from their effect by coating them.

One of these methods is coating the enzyme with a polymer structure. In this method, a nanocomposite network is created around each enzyme molecule by the polymerization process to prevent its destruction. These enzyme-containing nanoparticles last up to 5 months at a temperature of about 8 degrees Celsius. Enzyme coating in the food industry is one of the important processes for maintaining quality and improving food preservation.

7. Dairy Industry

Types of Nanoparticles Used in Dairy Industry

- ✓ **Inorganic nanoparticles:** Such as silver nanoparticles, zinc oxide, titanium dioxide, and nano calcium, which are used to increase shelf life, improve packaging, and enrich dairy products in the dairy industry.
- ✓ **Organic nanoparticles:** Such as nano casein, nano emulsions, and nano carbohydrates, which are used to improve texture, increase nutritional value, and reduce waste.
- ✓ **Nano liposomes:** Used to increase the bioavailability of vitamins, omega-3 fatty acids, and bioactive compounds.

Figure 5. Nanotechnology in Food Science: Food Safety, Food Packaging and Food Processing

Nanotechnologies in Dairy Product Processing

1. Nano filtration for Milk Purification

- ✓ A membrane process used to remove bacteria, viruses, lactose, and undesirable compounds from milk without the use of heat.
- ✓ Increases the stability of milk without the need for intense pasteurization.
- ✓ Helping to isolate beneficial proteins such as casein to produce protein-enriched products.

2. Nano encapsulation to protect sensitive compounds

- ✓ Use of lipid and polymer nanoparticles to coat sensitive compounds such as vitamins, probiotics and minerals.
- ✓ Increase stability and bioavailability of nutrients in the body.
- ✓ Improve shelf life and taste quality in fortified dairy products.

3. Biological nanostructures to optimize fermentation and improve texture

- ✓ Use of nano enzymes to accelerate and improve the fermentation process in yogurt and cheese production.
- ✓ Create nanostructures (nano enzymes) in milk proteins to improve the texture and consistency of fermented products.
- ✓ Increase the growth of beneficial probiotic bacteria in dairy products.

4. Nano emulsions to increase solubility and absorption of nutrients

- ✓ A combination of fats and water at the nanoscale to improve the dissolution and absorption of fat-soluble nutrients such as vitamins A, D, E and K.
- ✓ Use in the formulation of low-fat products without compromising quality
- ✓ Prevents oxidation and spoilage of fats in dairy products.

5. Nano sensors for quality control and safety

- ✓ Nano sensors in smart packaging to detect signs of spoilage such as microbial growth or PH changes.
- ✓ Quality control of raw milk and dairy products during production.
- ✓ Increase food safety and reduce waste of dairy products.

6. Nano-bio composites in dairy product packaging

- ✓ Use of nanoparticles such as silver, copper, zinc oxide and nano-clay in packaging to reduce microbial growth in packaging
- ✓ Reduce oxygen and moisture penetration to increase the shelf life of dairy products
- ✓ Production of biodegradable and environmentally friendly packaging.

Application of nanoparticles in dairy industries

The use of nanoparticles in dairy industries can lead to improved quality, increased shelf life, improved safety and increased nutritional value of products. The following are the main applications of nanoparticles in this industry:

1. Increased shelf life and improved packaging of dairy products

A) Antibacterial and penetration-resistant nano-packaging

Nanoparticles such as silver (Ag), copper (Cu) and zinc oxide (ZnO) have antimicrobial properties and can prevent the growth of bacteria and fungi in packaging. The use of these nanoparticles in packaging films leads to increased shelf life of dairy products such as milk, yogurt, cheese, and cream.

B) Nano-packages with high mechanical properties and impermeability

Nanoparticles such as nano-clay and nano-cellulose increase mechanical strength and reduce oxygen and moisture penetration into the package. This feature prevents the oxidation of fats in dairy products and prevents premature spoilage of dairy products.

2. Improving the quality and safety of dairy products

A) Removing contaminants and toxins from milk and dairy products

Some nanoparticles such as nano-titanium dioxide (TiO_2) and iron nanoparticles (Fe_2O_3) can act as adsorbents and remove toxins, heavy metals, and pesticide residues from milk and other dairy products. This leads to the production of healthier and safer products.

B) Increasing the stability of proteins and nutrients

Some nanoparticles, such as nano casein, can increase the stability of proteins and prevent their degradation under adverse temperature or acidic conditions. This technology is very useful and important in the production of protein-enriched yogurt and milk.

3. Improving nutritional properties and increasing the bioavailability of nutrients

A) Nano encapsulation of nutrients

Nano encapsulation is one of the new technologies in the dairy industry that increases the stability, absorbability and controlled release of beneficial compounds such as vitamins, minerals and probiotics. Some of the applications of this technology include the following:

- ✓ Nano encapsulation of vitamin D and calcium to increase their absorption in the body.
- ✓ Protection of probiotics (beneficial bacteria) in yogurt and buttermilk against adverse digestive conditions by nano encapsulation.

B) Increasing the solubility and bioavailability of beneficial compounds

Some nanoparticles, such as nano emulsions, improve the solubility of lipophilic compounds such as vitamins A, D, E, and K, which are usually poorly absorbed in an aqueous environment. This method helps to enrich dairy products with fat-soluble vitamins.

4. Optimization of production processes and quality control

A) Nano sensors for detecting spoilage of dairy products

Nano sensors can be used in smart packaging and, by detecting chemical and biological changes, indicate signs of spoilage. This technology increases consumer safety and reduces food waste.

B) Optimization of fermentation processes and shelf life of yogurt and cheese

The use of nano enzymes in the production process of fermented products such as yogurt and cheese increases efficiency, improves the texture and taste of products, and increases their shelf life.

Impact of nanoparticles on consumer health

The use of nanoparticles in the food industry, including the dairy industry, can have different impacts on consumer health. The rate of absorption and excretion of nanoparticles in the body depends on various factors such as the type of nanoparticle, size, shape, surface coating and route of entry into the body. Some nanoparticles, such as nano calcium and nano liposomes containing vitamins, are easily absorbed and metabolized, while others, such as metal nanoparticles (silver, zinc oxide, titanium dioxide), may accumulate in the body and cause unwanted effects. Studies have shown that a small amount of these nanoparticles is absorbed through the digestive tract and a large part of them is excreted through the kidneys and liver, but in some cases there is a possibility of accumulation in tissues.

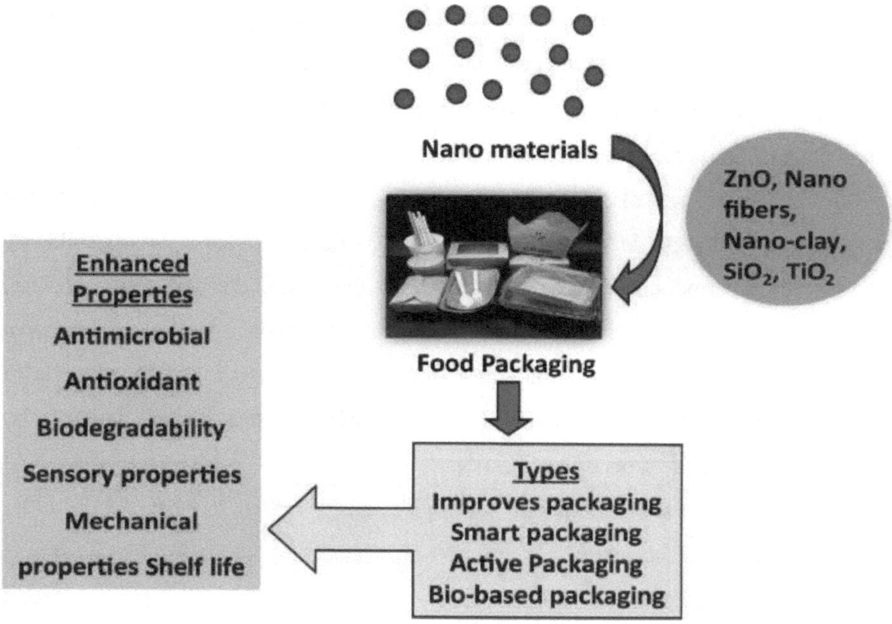

Figure 6. The intervention of nanotechnology in food packaging

Potential risks from excessive consumption of nanoparticles include cellular inflammation, oxidative stress, organ dysfunction and even Geno toxicity. Some studies have shown that nanoparticles such as titanium dioxide (TiO_2) in high doses may damage the intestinal wall and disrupt the balance of the gut microbiome. However, the long-term safety of nanoparticles in food is still not completely clear and requires further studies.

Health organizations such as the FDA and EFSA are closely examining the possible effects of these compounds on human health and have set standards for their permissible levels of consumption. Therefore, although nanotechnology has several advantages in improving the quality of dairy products, its use must be carried out with caution and careful monitoring to avoid possible side effects.

Nano encapsulation Sensory quality of food

The sensory quality of food, for example, smell and taste, influences its consumption and is therefore very important. However, controlling and stabilizing sensory quality is difficult, especially during food production and storage. The key to overcoming these issues could be in encapsulating the flavor before use in food.

Nano encapsulation may be the most important technology in food science, capable of controlling the release of flavors and protecting against flavor degradation during processing and storage. Saliva can activate nano capsules to release flavors suddenly or over a long period, ensuring delivery at the desired time and speed. This technique also offers a targeted delivery system to increase the bioavailability of bioactive compounds such as vitamins, lipids and flavonoids. These compounds struggle to survive in the highly acidic environment of the stomach, where they are degraded by enzymes. Here are the roles of nano carriers:

- ✓ Protecting these compounds;
- ✓ Increasing surface area;
- ✓ Increasing solubility;
- ✓ Increasing bioavailability.

What are the characteristics of a suitable nano delivery system?

A system should be chosen that can be easily incorporated without interfering with the texture and taste of the food. Particle size is important; It should be able to deliver the active compound precisely to the target, ensure that the target is available at the right time and at the right rate, and be efficient enough to retain the active compound for a long time. Nano encapsulation improves the shelf life of foods by slowing down the degradation process.

For example, edible nano coatings can provide a barrier to moisture and gas exchange in addition to providing colors, flavors, antioxidants, and anti-browning agents. This increases the shelf life of the produced foods, even after their packaging is opened.

Application of Nano sensors in Food Packaging

Packaging plays a crucial role in keeping our food fresh and healthy. The use of nanotechnology in food packaging has the following applications:

- ✓ Improving food protection mechanisms;
- ✓ Ensuring that the product remains safe until consumption;
- ✓ Extending the shelf life of the food product.

In addition to being strong and biodegradable, the materials used in packaging must also be gas and moisture permeable. Nanocomposites and nano laminates are actively used in packaging to provide a barrier against severe thermal and mechanical shocks and increase the shelf life of food.

Smart or active nano-based packaging offers increased strength and improved impermeability properties with the potential to improve food safety. Nano sensors can be used in smart packaging for rapid, sensitive and reliable detection of microbes, pathogens, harmful chemicals and pesticides. For example, Kraft Foods has developed a highly sensitive nano sensor that detects food gases released by food as they are released. A sensor strip indicates freshness by changing the color of a clear optical signal.

The Connection Between Health and Nanotechnology in Food Processing

There are many benefits to using nanotechnology in food and its packaging that have not yet been widely used in this field. Researchers are finding new ways to enhance food quality while maintaining the nutritional value of the final product. However, growing concerns about food quality and health benefits, public safety, and environmental concerns must also be considered.

Nanotechnology in the Food Industry

The unfamiliar properties of nanomaterials can pose unpredictable threats, the main concern being their migration from packaging and bioaccumulation in humans. These concerns can be addressed by standardized testing to reduce consumer concerns and improve the acceptability of such foods. Food regulations serve to regulate consumer health issues.

Nanotechnology in the food industry and quality sensor capsules

Food quality sensors – for example, in the area of smell and taste – affect consumption and are therefore of great importance. It is noted that the issue of quality sensors and their stability and stabilization is a complex process, especially in the production and storage of products. The key to solving this problem is the encapsulation of flavors before use in food. Nano capsules are probably the most important technology in food science that can control the release of flavors. They can also protect flavors from quality loss during the production and storage process.

Nano Food Capsules

Properties of Nano Capsules

Saliva can activate nano capsules to release the flavor of food explosively. They can also ensure the flavors at the right time and amount over a fixed period. This technique can be used to optimize the delivery of a specific amount of active ingredients needed by the body. These include substances such as vitamins, lipids and flavonoids. In fact, these substances must remain in the completely acidic environment of the stomach and not be subjected to enzymatic decomposition.

These nano carriers can provide protection, increase the surface area, increase solubility and provide stable shelf life of the substances. However, a system that can be used easily and without interfering with the texture and taste of food must be selected and designed. Particle size is important. These particles must be able to deliver the active ingredients precisely and at the right time and ratio to our target. They must also be able to maintain the active ingredients we want for a long time with the same quality. Nano capsules improve the shelf life of food by reducing the process of degradation.

For example, edible nano-coatings, in addition to the proper transfer of color, flavor, antioxidants, and anti-discoloration agents, can create a barrier to unwanted moisture and vapors and gases resulting from food spoilage. In this way, nanotechnology in the food industry helps to increase the shelf life of food, and this can be extended to after opening the food.

Food Packaging Industries

Nano Sensors and Packaging

Packaging industries play an important role in keeping food fresh. Food industry nanotechnology is improving food protection mechanisms every day. This knowledge can also create confidence in food safety and increase shelf life. The materials used in packaging must increase their resistance and reduce the biodegradability of the materials.

They can also prevent the penetration of various vapors and moisture into the food. For this reason, the use of materials such as nanocomposites and nano laminates is done in packaging. This will create a barrier to severe thermal and impact shocks and increase the shelf life of food. Smart or active packaging based on nanotechnology includes greater strength and impermeability and is considered as a measure of food safety. Nano sensors can be used in smart packaging. As a result, it is possible to quickly, accurately and reliably find out about microbes, pathogens, harmful chemicals and even toxins remaining in food.

Food packaging spoilage sensors

For example, a reputable food company has been able to detect the food spoilage gas that is generated in itself by developing nanotechnology in the food industry. In fact, with the help of a colored sensor strip, it is possible to determine the freshness of the food in front of you.

Nanotechnology in food vs. food safety

All of the above are numerous advantages that can be applied with the help of nanotechnology in the food and packaging industries. Meanwhile, researchers are discovering new ways to improve the quality of foods. Also, nutritional values are fully preserved until the end of the shelf life. However, amid the increasing attention to food quality and the aforementioned health benefits, it is necessary to also consider public safety and environmental concerns.

The unknown properties of nanoparticles can also pose unforeseen threats. For example, the issue of the transfer of properties of nanoparticles from packaging to food. Bioaccumulation of these materials in the human body is another issue that can be mentioned.

Food processing

Food processing is the conversion of raw materials into food and other forms of it through marketability and long shelf life. Processing includes the incorporation of nutrients, mineral and vitamin enrichment, gelling and viscosifying agents, nutrient delivery, removal of toxins, prevention of pathogens, preservation and improvement of food consistency that is better for marketing and distribution. Processed foods are usually less susceptible to premature spoilage than fresh foods and are more suitable for long-distance transportation from source to consumer.

All of this has become more effective today with the integration of nanotechnology. Nanoparticles are added to many foods to improve flow properties, color, and stability during processing or to increase shelf life. Nano capsule delivery systems play an important role in the processing sector and behavioral properties are maintained by encapsulating simple solutions, colloids, emulsions, biopolymers, etc. inside food. Self-assembled structural lipids at nano size act as a liquid carrier of healthy water-insoluble components and fats called nano droplets. They are used to prevent the transfer of cholesterol from the digestive tract to the bloodstream.

Nano encapsulation

Encapsulation is a technique by which sensitive materials are encapsulated within a coating. The coating material protects the sensitive materials from adverse reactions and controls the release of the materials. In addition, the encapsulation process can convert liquids into powders that are easy to handle. Nano encapsulation has several advantages, including increased stability, ease of transportation, retention of volatiles, protection against oxidation, PH and moisture controlled release, flavor enhancement, sequential delivery of multiple active ingredients, modification of taste, and increased bioavailability and efficacy.

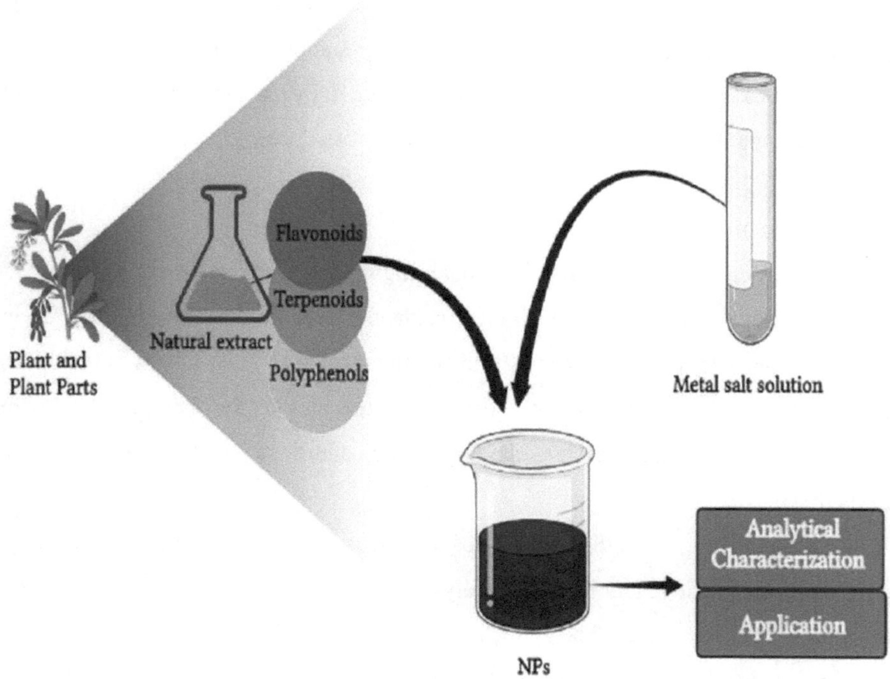

Figure 7. Recent insights into metallic nanoparticles in shelf-life extension of Agrifoods

Nano capsules are prepared basically by six methods namely nanoprecipitation, emulsion diffusion, double emulsion, emulsion shrinkage, polymer coating, and layer-by-layer. Nano capsules are used to lipophilize health supplements such as vitamins and minerals in food, fatty acids, and growth hormones, increasing the nutrient content of food. Nano encapsulation can be used to protect sensitive bioactive substances such as omega-3 fatty acids, carotenes, vitamins, coenzyme Q10, polyphenols, etc. from harmful environmental conditions during processing and delivery.

Nano carriers that can be used as coatings with numerous advantages include nanofibers and nano hydrogels. Silica nanoparticles, for example, are used as anti-caking agents in processed granular or powdered foods (such as tomato powder, table salt, and spices). These nanomaterials prevent aggregation and maintain flow ability. Silica nanoparticles are also used for targeted drug delivery and release due to their high biocompatibility and convenient formulation with drugs.

Nano emulsion

Nano emulsions for delivering functional ingredients are an emerging area of nanotechnology applied in the food industry. Nano emulsions consist of oil droplets in the nanometer size range, between 10 and 100 nm, dispersed in an aqueous continuous phase. The droplets surrounded by surfactant molecules of the nano emulsion can protect the flavor compounds from the production conditions and during the shelf life of the beverage.

It is claimed that nano emulsions can absorb flavor and protect it from temperature, oxidation, enzymatic reactions and hydrolysis, and are thermodynamically stable over a wide range of PH values. The application of nano emulsions in ice cream has made it possible to produce ice cream with a fat content of less than 16% actual to 1% without compromising the taste and flavor. Water-in-oil emulsions with a size of 10 to 500 nm have been used with the aim of achieving faster and easier melting by adding poly sor bates, etc. Micelles are claimed to help in the uniform melting of frozen foods in the microwave.

Food Packaging

The need for protection, resistance to tampering and specific physical, chemical or biological requirements of food are met by food packaging. Also, any nutritional information about the food consumed can be obtained with the label on the food package. Packaging is of great importance in preserving food to market. Innovations in packaging have led to quality packaging and a customer-friendly approach with high shelf life, biodegradability and many more.

Nanotechnology in packaging is categorized based on the purpose of application. Food packaging with the help of nanotechnology can be broadly divided into the following three main categories.

Improved packaging

Thereby, nanomaterials are mixed into the polymer matrix to improve the gas barrier properties, as well as the temperature and humidity resistance of the packaging.

Active packaging

Using nanomaterials to interact directly with food or the environment has been shown to better protect the product. For example, silver nanoparticles and silver coatings can have antimicrobial properties, and other materials are used as oxygen or ultraviolet absorbers.

Smart/intelligent packaging

This type of packaging is designed to detect biochemical or microbial changes in food, for example identifying specific pathogens growing in food, or specific gases resulting from food spoilage. Some "Smart" packaging is therefore also developed for use as a tracking device for food safety or to prevent fraud. Nanotechnologies used in food packaging are increasing worldwide, and many international food companies are exploring their potential applications.

Among the applications of nanotechnology for the food sector, nanotechnology-derived foods account for the largest share of the current and near-term forecast market. Some examples include enzymes combined with gold nanoparticles for microbial detection, gas sensing related to food product conditions such as nano fibrils of perylene-based fluorophores by detecting gaseous amines, indicating fish and meat spoilage. Others include nanocomposites of zinc oxide and titanium oxide for detection of volatile organic compounds. Nano barcodes are used for labeling as well as for security. Therefore, the use of smart sensors is beneficial for consumers in terms of better quality identification and for manufacturers for fast distribution and authentication of food products.

Antimicrobial Packaging

Barriers include natural nanoparticles to control microbial growth that leads to pathogens or spoilage. Silver nanoparticles are used in all forms including bio textiles, electrical appliances, refrigerators, kitchen utensils. Silver nanoparticles show the necessary action in bulk and its ions have the ability to inhibit a wide range of biological processes in bacteria. The antibacterial nature of zinc oxide increases with decreasing particle size, it can be excited by visible light and is incorporated into a number of polymers including polypropylene. Chitosan, a biopolymer derived from chitin, has recently been reported to have antimicrobial properties. Antimicrobial packaging will be a very healthy and customer-friendly product.

Biodegradable packaging

Pollution is the most important factor affecting environmental characteristics. The impact of non-degradable plastics changes the nature of the soil and the accumulation of toxic gases in the atmosphere leads to global warming, hence biodegradable plastics came into play but they lack mechanical strength and permeability to water and gas. Nanotechnology-incorporated packaging materials (made from natural or synthetic nanoparticles) overcome these disadvantages (low strength and permeability) and combine biodegradable and renewable properties with high mechanical strength.

Nanoparticles are used in the form of proteins, carbohydrates, lipids from animals and from plant materials, as well as metal oxides nanoparticles and carbon nanotubes.
In addition, collagen, cellulose from corn are synthesized in the form of nanofibers and are highly porous in nature. These nanomaterials are added to nano clays and used for convenient packaging. They also have additional new properties such as sensing, antibacterial function and are used as biocatalysts.
As a result, nanotechnology will serve as an important tool to overcome the existing challenges associated with packaging materials. This development will have a positive impact on the shelf life, quality, safety and security of food, which will ultimately benefit consumers and producers. However, the precautionary principle should be applied and further research is needed, especially on the migration behavior of nanomaterials in food and their potential effects on health/safety as well as the environment. A sustainable packaging solution can only be achieved if it is socially responsible, economically viable and environmentally sound.

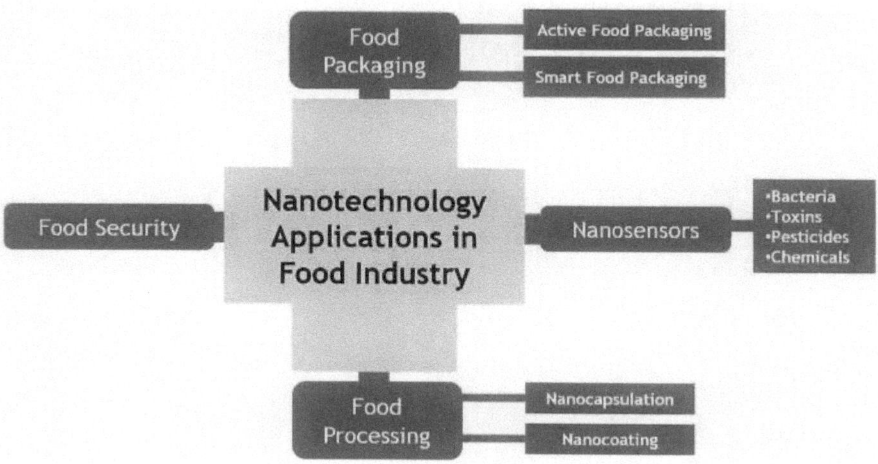

Figure 8. Application of Nanotechnology in the Food Industry

Chapter III

Application of Nanotechnology in Food Packaging

Nanotechnology is one of the best and most popular technologies that has been able to find its way into several different industries and is used in various fields. The packaging industry has also been affected by this technology, because it is one of the important industries in which health and safety are of particular importance. If food must be stored for a long time, many fungi and bacteria may penetrate into them.

But by using packaging, all materials can be protected against fungi and other damage. But what is the role of nanotechnology in the food industry? Nanotechnology plays an important role in this industry and increases the capabilities of packaging. For example, nanotechnology can help increase the resistance of the package to various factors such as harsh environmental conditions, impact, abrasion, preventing gas from entering the food, and increasing mechanical properties. In addition, the use of nanotechnology will help preserve the taste, color, and shelf life of food in the long term.

Since nanotechnology is a new technology in the world, it has more applications in the production of new materials. In addition to the food industry, nanotechnology is also used in other industries such as the environment, medicine, textiles, cosmetics, paints, drugs, and lubricants.

Food packaging with nanotechnology

Now that we are familiar with the use of nanotechnology in food packaging, it is better to get acquainted with how food is packaged using this technology. Packaging is a process that can protect the product from any damage and chemicals. Cheap and lightweight packaging should be used for food. Also, the transportation process should be made easier by using packaging. Packaging that is made using nanotechnology is so important that it has caused major changes in this industry.

Food packaging using nanotechnology is divided into several categories, which are as follows:

- ✓ **Smart product packaging**

This type of packaging includes several labels that indicate the health of the products at different times. In addition, RFID labels and various sensors will also be used on these packages. All labels and information on this type of packaging can be identified with the help of radio waves. Due to the use of RFID tags, it is also possible to fully monitor all products remotely. Smart packaging helps to increase the quality of food and keep products fresher, in addition, there is also the ability to absorb oxygen in these packages. Nanotechnology is also widely used to produce various polymers and create high coverage.

- ✓ **Active food packaging**

In this method of packaging, food is in contact with the packaging in two ways: directly and indirectly. In such conditions, the packaging has the ability to change the composition of the food; in fact, this packaging will create special conditions for the food, which will lead to high quality, increased shelf life, food safety, and mechanical properties. Antimicrobial packaging is one of the types of active packaging that prevents the penetration of any microbes into the food. Also, most active packaging uses silver nanoparticles that will eliminate the presence of any bacteria. The advantages of this packaging include increasing the shelf life of the food.

- ✓ **Improved food packaging**

Today, nutrition science has made extensive progress and humans have made many achievements in this field. But still, many dangers such as bacteria, infections, organisms, viruses, mold threaten human health. Since harmful substances can penetrate into food and endanger human life, the use of advanced or improved packaging is very important. Improved packaging is one of the most important packaging that has many polymer chains.

This packaging has many advantages, for example, it prevents gases from entering the packaging. It will also increase the resistance of food to environmental factors such as heat.

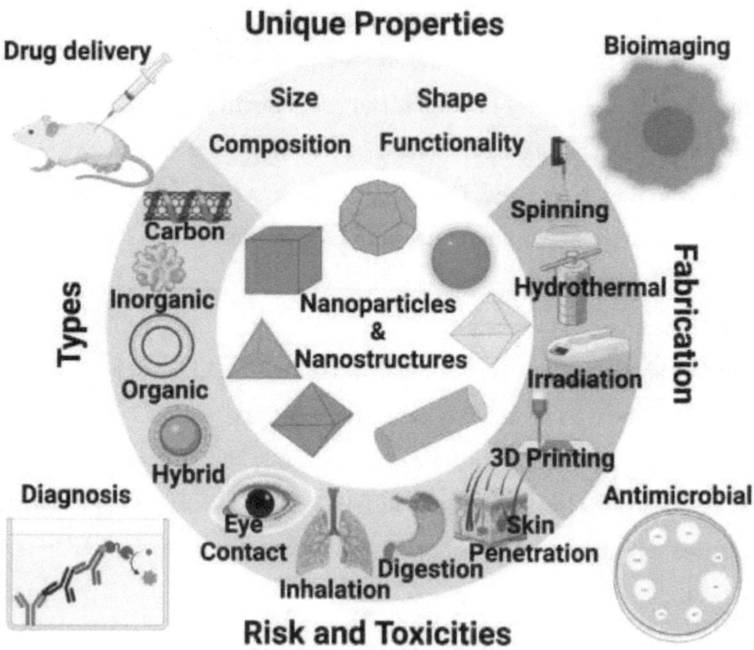

Figure 9. Nanoparticles and Nanostructured Materials: Bioimaging, Biosensing, Drug Delivery

Nanotechnology methods in food packaging

Nanotechnology has adopted various methods in food packaging, and in this section we will examine a few of them:

- ✓ **Nanocomposites**

Many polymers have been reinforced using nanoparticles, which are called nanocomposites. Using these particles, food packaging is better and more developed. The strength of nanocomposites is very high and increases the resistance of food in humid conditions. Various materials are used in the structure of this type of nano, clay nanoparticles are one of these materials, which increase the lifespan, mechanical resistance, prevent the penetration of moisture and materials, and prevent food spoilage.

- ✓ **Nano sensors**

Nano sensor is a very important option in food packaging. In addition to increasing safety, this nano is also a very important and cost-effective option. In the production of this type of nano, various materials are used, of which gold nanoparticles are one of the types. Nano sensor is used for many purposes. Also, using this method, all microbes, chemicals, bacteria and viruses can be identified.

- ✓ **Nano coatings**

This method has been used since the distant past and is still widely used in various industries today. The width of these Nanos is large and can be used for various products such as apples, cheese, French fries, vegetables, fruits, chocolate. Using this product, the resistance of food is increased, and it also makes the packaging more beautiful due to its beautiful appearance. This product is also used to create a strong coating against gases, toxic and harmful substances.

Packaging Innovation: Nanotechnology and Modern Technologies

In the field of food packaging, advances in nanotechnology have revolutionized the quality and shelf life of products. Combining nanoparticles with packaging films increases resistance to moisture and oxygen penetration and prevents rapid deterioration of products. In addition, the integration of nanotechnology with modern packaging systems such as vacuum sealers have enabled the creation of more precise and hygienic packaging; so that not only the shelf life of food is increased, but also their safety level is improved.

This new approach paves the way for the design of smart packaging that can quickly detect environmental changes or signs of spoilage and provide the necessary information to the consumer, thus ensuring the health and quality of food products.

Scientists have identified the potential for using this technology in almost all sectors of the food industry.

Two of the most important areas are food processing (improving the texture and quality of food) and food packaging. Of these, nanotechnology is most widely used in food packaging because it is not added directly to the food and the natural structure of the food is preserved. The use of nanotechnology in food packaging will protect food from pathogens and harmful gases.

Nano sensors are used to detect contamination and evaluate packaged materials, as well as protect food from ultraviolet radiation. This technology improves food packaging and its capabilities and can be used to detect bacteria in food packaging, which creates better taste and quality of food, increases safety and barrier properties against gas and moisture penetration, increases shelf life and prevents food spoilage. As additives that have been used in the food packaging industry, we can mention nanomaterials including silver nanoparticles, titanium nitride, zinc oxide and nano clay.

Types of nanotechnology methods in food packaging

Next, we will consider methods in food packaging using nanotechnology and examine which of these methods has what application in packaging and which is the most appropriate strategy for the packaging line.

1) Nanocomposites

Polymers reinforced with nanoparticles are called nanocomposites, with the use of which the food packaging industry has developed. Polymer nanocomposites have higher strength than pure polymers, high resistance to ignition, better thermal properties, low melting point and gas transition temperature, and good resistance to moisture conditions. In most cases, 5% by weight of clay nanoparticles have been used in the structure of nanocomposites.

The use of nano clay in creating nanocomposites leads to the creation of features such as high mechanical strength, low weight, improved barrier properties against the passage of liquids and gases, preventing the penetration of oxygen and moisture in food, and thus preventing food spoilage. The use of biocompatible nanocomposites in food packaging, in addition to preserving food, increasing the shelf life of food, has solved biological problems and reduced the need to use plastics, and has improved mechanical, abrasion, and thermal properties, which are important features in packaging materials.

Polymer nanocomposites or bio-nanocomposites are hybrid nanostructured materials with improved mechanical, thermal and gas properties that cause less environmental pollution due to the reduction of plastics in food packaging. These materials have greater potential and stability in active food packaging industries to maintain antimicrobial activity and reduce the transfer of metal ion derivatives to packaged foods.

2) Nano coatings

Nano coatings have been used for several decades and are now widely used. Nano coatings are layers that are a few nanometers wide, in fact, a nano coating is a wax coating that is significantly and widely used as a coating for food products such as cheese and apples. The main role of nano coatings is to protect food and also to beautify the packaged surfaces. Edible nano coatings have been used in various types of food such as meat, vegetables, fruits, cheese, candy, chocolate, French fries and bakery products. In general, nano-coatings are mostly used to prevent the passage of gases, moisture and fat.

Nano-coatings are antibacterial coatings that do not contain chemical germicides and are resistant to chemicals, alkalis and heat and are flexible. In fact, nano-coatings are thin layers that have organic or inorganic origins and their function is to prevent penetration and migration, which in the packaging industry increases the quantitative and qualitative characteristics of food products.

3) Nano-sensors

Bio-sensor technology is an important option in the agricultural and food industries to ensure food quality and safety with cost-effective and rapid methods of controlling products and processes and detecting gases in packaged foods for the integrity of packaged materials. In the manufacture of biosensors, various types of nanomaterials are used, of which gold nanoparticles are one.

These materials provide other advantages such as suitable size, low cost, fast response, accuracy and sensitivity. The purpose of using these materials in food packaging systems is to identify nanoparticles released in packaged food in a short time, other applications include improving the sensitivity and ability to detect chemical compounds, bacteria, viruses and toxic substances.

These materials can monitor the quality and freshness of food during storage and its hygienic conditions. Nano sensors react to environmental factors such as temperature, humidity, oxygen content, microbial agents and pathogens and provide useful information to manufacturers and consumers about the storage conditions of materials in terms of temperature, expiration date.

Nano packaging by smart packaging method

One of the capabilities of smart packaging materials is the ability to absorb oxygen from food, during which the oxygen present in the packaging is absorbed, thus stopping the growth of microbes and maintaining the taste and quality of the food. Other capabilities of smart packaging include the passage of carbon dioxide and the use of moisture-absorbing materials. Carbon dioxide can stop the growth of microbes in products such as red meat, poultry, and cheese.

The absorbent materials used in smart packaging are ethylene gas and antimicrobial materials. Nanotechnology is used in the manufacture of smart packaging to extend the life of the product, produce polymers with properties that prevent the penetration of gases and moisture, and create functional coatings. This system includes labels on the goods or products that are identified by radio waves. Using the RFID system, the status (safety and quality) of the product, package, or packaging environment can be monitored remotely.

Nano packaging by active packaging method

In this method, active materials that are directly or indirectly in contact with food have the ability to change the composition of the food or the atmosphere around it. Antimicrobial packaging is a type of active packaging in which the shelf life of the product is increased with the help of antimicrobial agents used in the preparation of the packaging material to prevent the growth and reduction of microorganisms. In this packaging, the use of nanomaterials such as copper oxide, silver, silver oxide, titanium oxide, magnesium oxide and carbon nanotubes that show antimicrobial properties in direct contact with food or its environment is common.

In the meantime, the use of silver nanoparticles in food packaging as an antibacterial agent has grown significantly. Active packaging, which has been developed mainly for antimicrobial applications, has created great hopes for improving the safety, quality of food and the shelf life of food products.

Nano packaging as an improved packaging method

Since the direct use of antimicrobial agents in foods can be harmful to the health of consumers, antimicrobial packaging has become very important. The most important features of this type of packaging are their increased ability to prevent the entry of gases, regulate the temperature, and resist moisture of food.

Therefore, to prevent factors of corruption such as those listed above, it is better to use this type of packaging. In order to maintain the quality of food in the packaging until it reaches the customer and to ensure that these packages do not harm the environment after use, it is better to use packaging that is produced based on nanotechnology; This type of packaging has created a huge and remarkable transformation in the food packaging industry, which can be divided into the following categories.

- ✓ **Smart Nanotechnology Packaging:** Smart nanotechnology packaging is a label that can show the freshness of a product over time; These packages work using radio waves, which can be remotely controlled using an RFID system and monitor the quality of the contents inside the package. Smart nanotechnology packaging allows the seller or even the consumer to be aware of the condition and status of the food inside the package; this type of Nano packaging has the ability to increase the efficiency of data transmission about the color, quality and freshness of the contents of the package during distribution.
- ✓ **Active nanotechnology packaging:** In this technique, active substances that are directly and indirectly in contact with food can change the composition of the food; Active nano packaging creates conditions in which the contents inside the packages are completely safe and their nature does not change under different conditions; These packages have antimicrobial properties and by preventing the growth of microbes and minimizing microorganisms, they increase the shelf life of food and maintain its quality and freshness.
- ✓ **Improved nanotechnology packaging:** With the many improvements and developments that have occurred in nutrition and food science, the risk of food contamination with various bacteria and microorganisms always threatens human health; Therefore, improved nanotechnology packaging should be used for food packaging; These packages, which have antimicrobial properties, do not endanger human health and are not in direct contact with food, and their use can increase the safety and quality of packaged food.

Nanomaterials used in food packaging

Nanoparticles are used in food packaging due to their very small size (less than 100 nanometers). These nanoparticles can improve the sensory and protective properties of products. For example, the use of antioxidant nanoparticles can prevent spoilage of food products and increase shelf life.

Nanoparticles can also act as nano sensors to detect spoilage or other indicators such as temperature and humidity. Although the use of nanoparticles brings many improvements in packaging, it also presents challenges. Some of the potential problems include issues related to human health, environmental impacts, and regulations related to the use of nanoparticles in consumer products.

Nanocomposites

Nanocomposites are formed by combining two or more different components, including a matrix and reinforcing nanomaterials. In food packaging, nanocomposites are used to enhance the structural strength of the material. These nanomaterials are usually made of materials such as silver nanoparticles or carbon nanomaterials and are added to food to increase the physical and mechanical strength of the packaging. The use of nanocomposites leads to improved packaging quality, increased shelf life and reduced food waste.

Some nanotechnologies in packaging

Nano sensors play an important role in food packaging. This technology allows for early detection of food spoilage. With the permission of nanoscale dimensions, these sensors examine food more accurately and sensitively. This early information allows producers and consumers to take the necessary measures to maintain the quality of food. Nano sensors also help in better time management of food consumption.

Nano coatings

The use of nano-coatings in food packaging is common in order to prevent the transfer of odor and oxygen to food. Due to their nano-sized dimensions, these coatings can be applied as a very thin and uniform layer on the surface of food. These features lead to an increase in the shelf life of food. Also, these coatings can help protect the contents of food from light, moisture and other environmental factors.

Humidity control

The use of nanomaterials in food packaging to precisely control humidity is one of the strengths of this technology. These materials act as the thinnest packaging layers and create less porosity in the packaging. This feature allows for better humidity control and prevents food spoilage due to humidity.

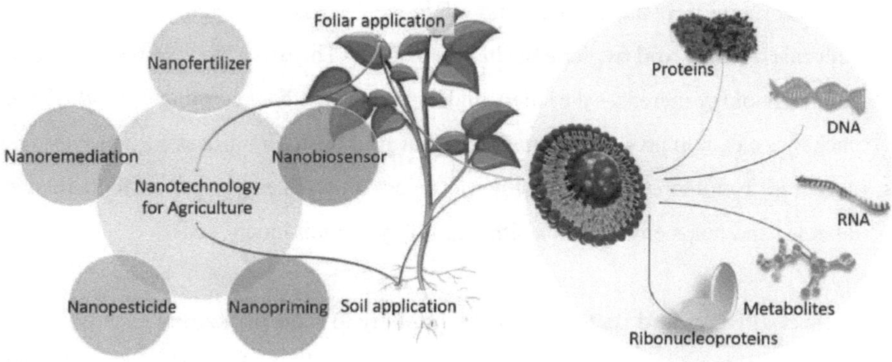

Figure 10. Regulation and safety measures for nanotechnology-based agri-products

Antibacterial and antiviral

Nanomaterials are used as antibacterial and antiviral materials in food packaging due to their size and ability to produce various nano-silicon structures. These features help increase food safety and hygiene and minimize the risk of infection transmission.

The role of nanomaterials in the manufacture of antibacterial and antiviral packaging

One of the important and innovative aspects of using nanotechnology in food packaging is the antibacterial and antiviral capabilities of these nanomaterials. Given the dimensions of nanomaterials and their surface properties, these materials can act as effective coatings on packaging and minimize the risks associated with bacteria and viruses.

- ✓ **Antibacterial packaging:** Using various nanoparticles, antibacterial layers can be applied to packaging surfaces. These layers can reduce the growth and spread of bacteria and act as a protective coating on food.
- ✓ **Antiviral packaging:** Nanomaterials are used as antiviral materials in packaging due to their size and ability to create nano-silica structures. These properties help increase food safety and hygiene and minimize the risk of infection transmission.
- ✓ **Increased safety and hygiene in the use of food:** The use of packaging containing nanotechnology increases the safety and hygiene in the consumption of food. These packages can help prevent the transfer of harmful bacteria, viruses and microbes to food and from there to the consumer. This action reduces the risk of food-related diseases and helps ensure the health and safety of individuals.

Possible challenges and issues of nanotechnology in food packaging

The use of nanotechnology in food packaging, with all its benefits, is also accompanied by challenges and issues that need to be considered.

- ✓ **Consumer safety and health:** Among the main issues in the use of nanotechnology in packaging is consumer safety and health. This is due to the possibility of nanoparticles attacking human body cells and the unknown side effects that may arise.
- ✓ **Nanoparticle stability:** The stability of nanoparticles over time and in different environmental conditions is a major challenge. Changes in the food environment and storage conditions can alter the properties of nanoparticles, thereby reducing the effectiveness of packaging.
- ✓ **Monitoring and standards:** Providing the necessary standards and monitoring to control the quality and safety of packaging containing nanomaterials presents its own challenges. Establishing accurate standards and monitoring methods that can assess the content of nanomaterials is crucial to ensuring consumer health.
- ✓ **Consumer information:** Consumers need accurate and transparent information about the use of nanomaterials in food packaging. Market communication and

consumer information based on sound and scientific information can help increase consumer confidence in this technology.

- ✓ **Environmental sustainability:** The environmental impacts of using nanomaterials are also an important issue to consider. The release of nanoparticles into the environment may have unintended consequences that need to be investigated and managed.

Nano plastics can be used in the production of bottles, plastic containers, and food packaging films, and due to their unique properties, they help to increase the shelf life and quality of food. One of the key features of nano plastics is increased resistance to oxygen and moisture penetration, which are among the main factors that cause food spoilage.

In addition, nanoparticles have antibacterial and antimicrobial properties that can prevent the growth of harmful microorganisms in food packaging. For example, silver nanoparticles, which have antibacterial properties, are used in food packaging to protect food from damage by microbes and bacteria. This property is especially useful for meat products and dairy products.

Another important application of nanotechnology in food packaging is to prevent the effects of light and heat on food. Some foods are sensitive to light and heat and may lose their quality due to these factors. Nano plastic packaging can prevent chemical changes in food by creating a protective layer against ultraviolet rays and heat, and increase their shelf life. Nanotechnology also helps reduce food waste. Nano plastic packaging can increase the shelf life of food and reduce food waste by preventing premature spoilage. This is not only economical for food producers, but also helps reduce food waste and protect the environment.

How nano plastics increase the shelf life of food

One of the biggest benefits of nanotechnology in food packaging is the increase in the shelf life of products. Nano plastics, due to their special properties, can act as a barrier against the penetration of harmful agents into food. These materials are particularly effective in preventing the penetration of oxygen and moisture, which are the main factors in food spoilage.

Due to the very small size of the nanoparticles, nano plastics can be transformed into a strong structure with high protective capabilities. This structure prevents oxygen molecules and other harmful gases from penetrating the packaging, and as a result, the oxidation process, which is one of the main reasons for the decline in food quality, is delayed. In addition, the use of nano plastics in plastic packaging can help prevent the growth of harmful microorganisms and increase the health of food.

Antibacterial nano plastics enhance the safety of packaged food

Another key benefit of nano plastics in food packaging is the use of nanoparticles with antibacterial properties. Silver nanoparticles and copper nanoparticles are among the materials used in nano-packaging due to their antibacterial properties. These particles are able to destroy harmful bacteria and microbes that may cause food spoilage, thereby increasing food safety.

Antibacterial nano plastics can help prevent the growth of microorganisms in packaging and help maintain the freshness and quality of food for a longer period of time. This feature is especially important in products sensitive to bacteria, such as meat products, dairy products and seafood.

Benefits of Nano plastics in Packaging Light- and Heat-Sensitive Foods

Light- and heat-sensitive foods such as oils, beverages, dairy products, and meats require packaging that can protect them from these factors. Nano plastics, as an advanced technology in packaging, offer significant benefits in protecting this type of food.

1. Protection against ultraviolet (UV) rays: One of the main benefits of nano plastics is the protection of food from harmful UV rays. These rays can lead to oxidation and chemical changes in food, and as a result, reduce their taste, color and nutritional value.

Nanoparticles used in plastics can create a protective layer that blocks the penetration of UV rays and prevents premature spoilage of food.

2. Reduction of heat transfer: Nano plastics help maintain the optimal temperature of food by reducing heat transfer into the packaging. Some foods may spoil faster due to heat due to their sensitivity to temperature changes. Nano-packaging is able to create a thermal insulation that prevents the temperature inside the package from increasing and prevents premature spoilage of heat-sensitive foods.

3. Longer freshness and quality preservation: Nano-plastics' protection from light and heat helps preserve the freshness, taste and nutritional value of food. Sensitive foods such as fruits and vegetables also benefit from the effects of nanotechnology. This technology helps preserve the natural color and texture of products and increases their shelf life.

4. Increase the shelf life of food: By creating additional protective layers against light and heat, nano-plastics can significantly increase the shelf life of sensitive foods. This feature is especially important for foods that are stored or transported under different temperature conditions. The use of nano-plastics in these cases helps reduce food waste and conserve resources.

5. Reduce the effects of oxidation: Light and heat can accelerate the oxidation process in food, resulting in reduced quality and spoilage of products. Nanoparticles used in nano plastic packaging act as a physical and chemical barrier and prevent this process from occurring. This is especially important in the packaging of oxidation-sensitive foods such as edible oils and processed products.

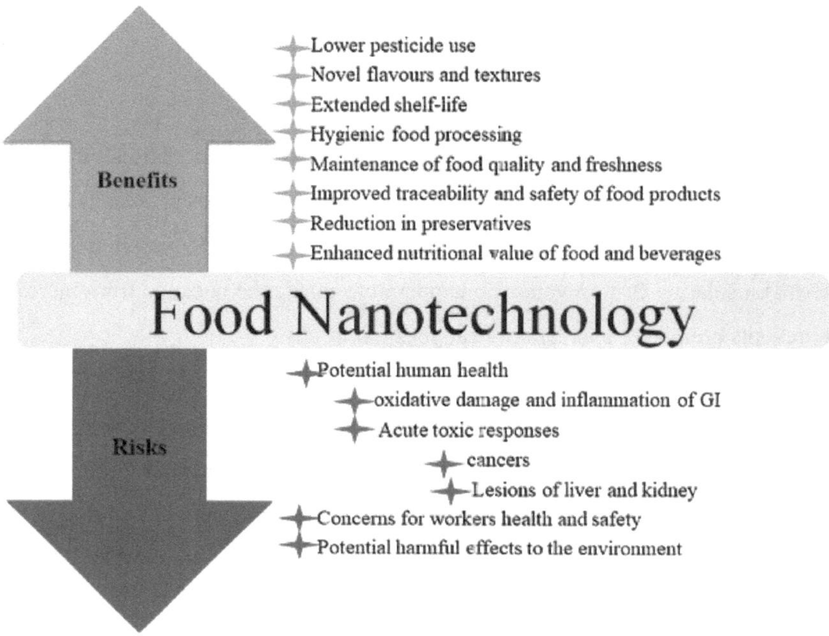

Figure 11. Nanotechnology, current uses and future applications in the food industry

How nanotechnology reduces food waste

1. Increase food shelf life: One of the most important factors that leads to food waste is its premature spoilage. Nanotechnology increases the shelf life of food through nano plastics and smart packaging. Nanoparticles such as nano silver or titanium oxide nanoparticles have antibacterial and antimicrobial properties that prevent the growth of bacteria and fungi in food packaging. This feature slows down the rate of food spoilage and increases its shelf life, which ultimately helps reduce waste.

2. Improve resistance to oxygen and moisture penetration: Nanomaterials can act as strong barriers to oxygen and moisture penetration in food packaging. Oxygen and moisture are two important factors in food spoilage that cause chemical changes and reduce the quality of products. By using nano-layers in plastic packaging, the entry of oxygen and moisture into the packaging is minimized. This process reduces oxidation and spoilage of food, especially in sensitive products such as oils, fruits and meat products.

3. Control the emission of gases inside the packaging: Some foods, such as fruits and vegetables, emit gases such as ethylene during storage, which accelerates the ripening process and ultimately premature spoilage. Nano filters and adsorbent nanoparticles in nanotechnology packaging can absorb these gases and control the rate of spoilage of products. This helps to increase the shelf life of food and prevent the wastage of agricultural products.

5. Improving temperature control in the supply chain: In the food supply chain, temperature control is one of the key factors in maintaining the quality of food. Nano-packaging using nanoparticles can help improve thermal insulation and prevent unwanted temperature changes during transportation and storage. By maintaining the appropriate temperature, food spoilage is reduced and their shelf life is increased.

6. Reducing the consumption of plastic materials in packaging production: The use of nanotechnology allows the production of lighter and less bulky packaging that, with the same level of food protection, prevents excessive consumption of plastic materials. This reduction in the use of packaging materials not only helps the environment, but also reduces production and transportation costs.
Advances in smart packaging to extend the shelf life of food products are a goal of many companies. These packaging systems will be able to repair small tears and holes according to environmental conditions (such as temperature and humidity changes) and inform the consumer of food spoilage. Nanotechnology can be useful in areas such as increasing the resistance to penetration of coatings, increasing wall properties (mechanical, thermal, chemical and microbial), increasing heat resistance, developing active antimicrobials and antifungal surfaces.

The financial prospects of nanotechnology show the packaging industry to be booming. The market share of this industry is currently around $ 1.1 billion and is expected to reach $ 7.3 billion by 2010. However, the smart packaging industry has gone ahead of what was predicted and the signs of its evolution are clearly visible.

Research by financial firm Frost and Sullivan found that consumer interest in healthy, fresh food in convenient packaging is driving the industry forward. There are many organizations working on smart packaging systems, including food manufacturer Kraft, which is working with Rutgers University on an electronic tongue project to add it to packaging.

This type of packaging contains an array of nano sensors that are highly sensitive to gases released by food and cause it to spoil, and change color, which is a clear sign of health or spoilage. Bayer Polymer has developed a plastic bag called Durethan KU2-2601 that is lighter and stronger than products on the market. It also has greater resistance to heat. The primary purpose of producing food packaging plastics is to prevent the contents from drying out and protect them from moisture and oxygen. The new coating is rich in silicate nanoparticles.

These nanoparticles largely block the penetration of oxygen, other gases and moisture, delaying food spoilage. Other organizations are finding ways to detect food spoilage using nanotechnology. For example, Agro Micron has developed a nano bioluminescent detection nozzle that contains a luminescent protein. In this design, the nozzle coats the surface of microbes such as Salmonella and E. coli, emitting light, and in this way food spoilage is detected.

The company hopes to bring the product to market under the name Bio Mark. The company is currently developing nozzles with new methods so that they can be used in maritime transport. In a similar strategy, to ensure food safety, EU researchers in the Good Food project are using portable nano sensors to detect harmful chemicals, pathogens and toxins in food. This would eliminate the need to send food samples to laboratories to determine the health and quality of products in farms and slaughterhouses.

The project is also developing the use of DNA biochips to detect pathogens. This method could be effective in detecting harmful and different bacteria present in meat or fish or fruit fungi. The project intends to develop filamentous micro sensors that can detect pesticides on fruits and vegetables as well as monitor environmental conditions in the fields. The innovation is called Good Food Sensors.

The EU-funded Bio Finger project, which aims to develop inexpensive, easy-to-use tools for environmental health, is also working on food analysis. In the cantilever-based devices, the tip is coated with a chemical and emits a signal when it encounters specific molecules. The Bio Finger is made smaller and more portable by using these cantilevers, which are attached to a microchip.

The US military is developing advanced sensors to be used against food attackers. Current systems take days to detect pathogens in food. Rapid pathogen detection by these sensors could soon make the technology widespread in the food industry. Researchers at the University of Bonn are developing anti-pollution coatings for packaging using the lotus effect (water droplets slide down the surface of a lotus leaf, forming nanoscale wax pyramids that cover the leaf surface). Slaughterhouses and meat processing plants could also use this technology. A research team from the University of Leeds in England found that nanoparticles of magnesium oxide and zinc oxide kill microorganisms. These materials are much cheaper to use than silver nanoparticles and could have many applications in food packaging.

Radio frequency identification (RFID) technology was developed more than 50 years ago, but today it has found its way into grocery stores. In this technology, data can be sent to wireless receivers using microprocessors. Today, this method can be used to track food items from the warehouse to the consumer. Unlike barcodes, which require manual scanning and reading one by one, RFID tags do not need to be read linearly and can read many of them in a second.

Chain stores such as Wal-Mart, Home Depot, Metro Group and Tsco are experimenting with the technology. The main disadvantage of this method is the increased production cost, which is a result of its silicon construction. By combining nanotechnology and electronics (nano tronics), these tags are cheaper and more efficient, and they are also easier to implement. A group of scientists from Northern Europe have formed the Nano food Consortium with the aim of developing applications of nanotechnology in this industry, with an emphasis on safe and healthy food.

The consortium consists of Arla Foods, Danisco A/S, Aarhus United A/S, Danish Crown amba and the Interdisciplinary Centre for Nanoscience. With an emphasis on providing consumers with healthy food, the consortium's priorities include developing sensors that can rapidly detect toxins in food samples or harmful bacteria, developing antibacterial surfaces for machines used in food production, developing stronger and cheaper coatings, and producing food with healthier food ingredients. The Danish Research Centre for Advanced Food Research (LMC), which is a consortium of Danish institutions active in the field of food science, announces its programs within the framework of its seventh program as follows:

- ✓ Basic understanding of food and animal nutrition for smart innovation;
- ✓ Systems biology in food research;
- ✓ Bio revision in the food sector;
- ✓ Technological developments;
- ✓ Food science;
- ✓ Customer-driven innovations and food communication.

They believe that focusing on these programs can lead to a comprehensive and multifaceted approach to food research and development in Europe. They also hope to use nanomaterials with functional properties for use in nano sensors and nano fluidic technology in the food industry. Advances in smart packaging materials will allow us to control the conditions of products during transport and use bio-based packaging methods.

Food processing

In addition to packaging, nanotechnology has a great impact on the development of functional and interactive foods; materials that respond to the needs of the body and can be effective in delivering nutrients. Various research groups are working on the development of new food materials on demand.

These materials remain inactive in the body and deliver nutrients to cells when needed. A key element in this area is the development of nano capsules that can be used in food to do the job of delivery well. Another development in food processing is the addition of nanoparticles to foods to increase their absorption in the body. One of the best bakeries in Western Australia has been successful in using nano capsules containing tuna oil (a rich source of omega-3 fatty acids). The company uses the nano capsules in its best-selling bread, Tip-Top, and the particles only break open and break down when they enter the stomach, thus avoiding the unpleasant taste of fish oil.

Nutralease, a company in the occupied West Bank, uses nanoscale self-assembled liquid (NSSL) technology to deliver nutrients.

The particles are in the form of micelles (hollow spheres made of fat and water) with a diameter of about 30 nanometers. Nutraceuticals have a watery interior and can be used to carry substances such as lycopene, beta-carotene, lutein, phytosterols, CoQ10 and DHA/EPA. These particles allow the compounds to easily pass through the stomach into the blood vessels. This increases their bioavailability. This technology is currently being marketed by Shemen Industries for the delivery of active Canola oil. The company claims it can reduce cholesterol absorption in the gallbladder by up to 14 percent.

A number of chemical companies are researching additives that are easily absorbed by the body and could extend the shelf life of products. The International Organization for Bio Delivery Sciences is developing nano snails with 50-nanometer spirals that could be used to deliver substances such as vitamins, lycopene and omega-3 fatty acids to cells without affecting the taste or color of the food. Kraft Foods has formed a group of researchers from 15 different universities to study the use of nanotechnology in food. This allows consumers to choose between different colors and flavors. The group is also working on developing smart foods with nano sensors, which release nutrients gradually.

These nano capsules are mixed with food but remain inactive until the right time. All these new developments are bringing the concept of whole foods closer to reality, and are expected to bring additional benefits in terms of energy, diagnostic functions, improved safety applications, and the development of anti-aging products for consumers. Nanotechnology is currently being used in the cosmetic industry, such as in the manufacture of clear creams.

Royal Body Care, a company that uses food nanotechnology, has launched a new product called Nano Ceuticals, which is an emulsion of particles less than 5 nanometers in diameter. The company claims that the product scavenges free radicals, increases hydration, and regulates the body's PH. The company is also developing nanoclusters and nanoscale powders that can be incorporated into dietary supplements and, when consumed, increase the absorption of nutrients in the body.

Food and cosmetics companies are working together to find new ways to deliver vitamins and get them directly into the skin. For example, Nestle, which owns 49 percent of L'Oreal, is developing a clear sunscreen that delivers vitamin E directly to the skin. The goal is to create a cream that is absorbed by the skin and releases vitamin E slowly, while also providing UV protection. Clear UV-blocking creams are already on the market, and L'Oreal expects this cream to take over the market with more applications. Other competitors, such as Estee Lauder, are developing anti-aging formulas that are made from nanoparticles.

The American company Oil fresh has launched a new nano ceramic product that can cut oil consumption in restaurants and ready-to-eat foods by half. As a result of this major change, products are prevented from oxidizing due to fats in the oil. Another thing is that the oil heats up faster, reducing the energy required for cooking.

Recently, Wageningen University in the Netherlands has established a research center that is working on the application of nanotechnology in the food industry. The Wageningen Biotechnology Center focuses on a variety of topics, including food quality and safety detection, food coating and delivery, micro- and nano-tools for chemical and physical processing, chemical biology, nano toxicology; technology research and customer science.

The German company Aquanova is developing a new technology that combines two active ingredients and uses them to reduce fat through nano carriers (hollow spheres with a diameter of 30 nanometers). This innovation could be a new breakthrough in weight control. Nova SOL Sustain uses CoQ10 to reduce fat and alpha-lipoic acid to suppress hunger. The technology is also used to produce vitamins such as SoluE, which is a vitamin E class, and SoluC, which is a vitamin C class.

In a different strategy, Unilever is producing low-fat ice cream by reducing the emulsion particles. It hopes that the use of these particles will reduce the fat content by up to 16 percent. The Woodrow Wilson International Center, a US-based scholarship foundation, has compiled a database of nanotechnology market customers and will soon announce 15 cases that are directly related to the food industry.

The list includes nano cetels produced by RBC Life Science, Shemen Industries' active canola oil, and silver nanoparticles used in LG refrigerators. Today, many countries around the world have realized the potential of nanotechnology in the food industry and are investing significantly in this direction. The UK Food Standards Agency (FSA) has organized research to achieve the potential of using nanotechnology in food, and specifically in food packaging.

At the same time, the government has allocated more funds for research and development in the field of functional foods, food delivery systems, and methods for optimizing food appearance such as color, taste, and consistency.

As nanotechnology's impact on the food industry increases and these products enter the consumer market, the importance of food safety becomes more important. This need will strengthen the adoption of nanotechnology in sensory applications, and in this way, food safety can be understood. Such as a type of technology that warns buyers and sellers that the expiration date of food is approaching. New antimicrobial coatings and plastic bags that repel contamination have made significant progress in ensuring the safety and security of packaged foods.

Although much attention has been paid to the application of nanotechnology in the food industry and products on the market, there are still many untapped capabilities, such as those previously discussed in the discussion of genetic modification.

The UK's Institute of Food Science and Technology has said in a report that more reliable data is needed before nanoparticles can be added to food. The report notes that current legislation does not require companies to label products containing nanoparticles, so it is unlikely that consumers will be able to tell whether they are in their food. It says that the effects of particle size should be considered in assessing the healthiness of these foods, as well as the type of ingredients.

The ETC group has also accused some major companies and active universities of trying to monopolize new foods (through patents), which could create barriers for many innovative companies in developing countries. Eventually, we will be able to make food from atomic and molecular compounds, which is called molecular food production. Some research groups are currently investigating this, but it is still a top-down approach, using cells rather than molecules. Although practical use of this technology is possible in the distant future, it is expected that this development can pave the way for the expansion of more efficient and sustainable food processing, which will consume fewer raw materials and produce higher quality foods.

Nanotechnology in food packaging in industries and companies

Currently, many companies such as Unilever, Keystone, Hershey, Food, nestle are working on nano-foods. It has been reported that Nestle and Unilever have discovered emulsions of nanoparticles that make the texture of food more uniform, and they can be used in the production of products such as ice cream. Other projects of this company are working on nano capsules containing enriched foods that gradually deliver nutrients and antioxidants to specific parts of the body.

This technology converts old food into nano-sized particles that are released inside the body and are well absorbed. This technology will have many applications in new foods. Another pioneer in the development of nano foods is Kraft, which took the first steps in the introduction of nanotechnology into the food industry by establishing the Nano Tek Consortium in 2000.

This consortium is a group of 15 universities and national research laboratories and is mainly active in the field of providing a variety of interactive foods and beverage products that adapt to the taste and individual needs of the consumer, covering a wide range of colors from color-changing drinks to new foods that adapt to the consumer's sensitivities (or nutritional needs). Another activity of this company is the production of nano filters that screen molecules more by shape than by size, which makes it possible to separate specific components of a product, even in the hands of the consumer.

Another goal of this company is to work on smart food packaging. Nano sensors that are sensitive to the release of chemicals caused by food spoilage can be used in smart packaging, so that as soon as the food begins to spoil, the color of the packaging changes, warning the customer. This system is much more accurate and reliable than selling by the expiration date.

Another company active in the nano food field is Nutralease, which is researching the technology of fortified foods and using nano capsules to increase the bio delivery of nutrients. This technology has been applied to a type of cooking oil that uses plant sterols to reduce cholesterol absorption and reduce the risk of heart disease. According to one report, this product reduces LDL cholesterol by about 14 percent.

Oil Fresh uses nano ceramics in its restaurant pans, which reduces frying time and oil consumption. The product allows restaurants to use vegetable oils instead of hydrogenated oils, which reduces trans fats and makes for healthier food. Another company, Voridian, has used Impern nanocomposites to make plastic beverage bottles. Impern is a type of plastic that is mixed with clay nanoparticles to create a plastic that is as hard as glass and stronger than glass, but less brittle than glass. The nanoparticle layer is designed to prevent carbon dioxide molecules from escaping from the beverage and oxygen molecules from entering the beverage, thus preserving freshness and increasing shelf life. Another company in this field is Nano cor. The company is the leading producer of plastic nanocomposites.

These plastics have unique properties, including better barriers to oxygen and carbon dioxide flow, which increases the shelf life of durable nanocomposite products. They also prevent odors from spreading and prevent the absorption of flavors or vitamins in food by the packaging. In general, the molecular design of these plastics is such that they improve the resistance of products to fire and the stability of their structure against heat. For example, these materials are used in baskets for boiling food and packages for use in microwaves.

Plastic nanocomposites can also be used in new food packaging. Another key product is electronic olfactory sensors (electronic noses). These devices mimic the human tongue and nose, except that they are more sensitive to subtle tastes and odors. An electronic nose is an array of nanoscale gas sensors, and the high surface area of the nanoparticles allows for the absorption of as much gas as possible. This technology, combined with pattern recognition technology, allows for the creation of a digital fingerprint of any specific odor.

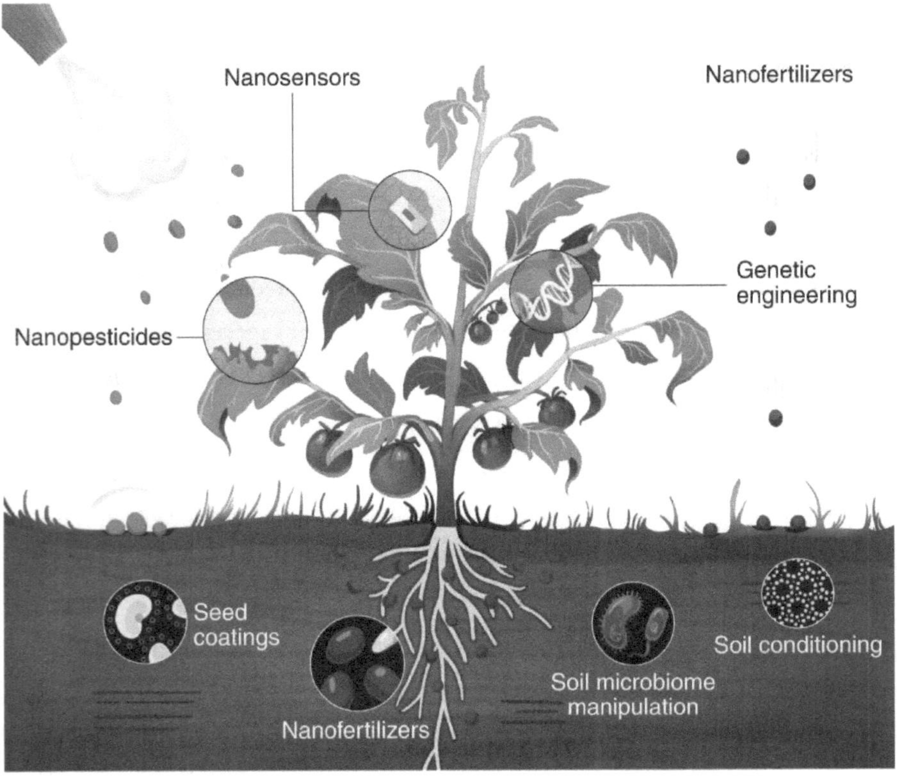

Figure 12. Technology readiness and overcoming barriers to sustainably implement

These products have been used in laboratories such as NASA to detect trace amounts of chemicals, but are now being used in the food industry to control the best levels of food produced.

They are also effective in detecting contaminants and in qualitative and general analysis of food. Some companies are currently using a type of electronic tongue that includes an array of liquid sensors (electrodes coated with conductive polymers) along with pattern recognition technology that is capable of distinguishing specific flavors from each other.

One of the important applications of this language is the taste test of beverages such as fruit juice, milk, coffee, mineral water and soft drinks, as well as the ability to taste chemicals at the PPT level, and its production cost is about 50 cents. This language will certainly find its vital role in food studies. The taste sensor in meat packaging is able to detect the first signs of food spoilage and warns of food spoilage by changing color.

Another type of sensor technology is nano barcodes, developed by Nano Plex Technologies. Nano barcodes are molecular versions of traditional barcodes and consist of metallic nanoparticles that have a unique, identifiable chemical fingerprint that can be read by a machine (likely a UV light with a light microscope). These types of barcodes can be used to protect brands and evaluate foods that would not normally be labeled with traditional barcodes.

They can also be used to detect pathogens in food, such as E. coli. In fact, pathogen detection is another major goal of nanotechnology in the food industry. Hong has also worked on biological Nano sensors.

These sensors can detect minute amounts of pathogens in food. They could also be used in food storage and transportation facilities for precise control at the molecular scale. He has also worked on so-called functional foods, exploring the role of nutrients in promoting health and preventing disease.

"Many foods have inherent disease-preventing properties, such as green tea, grape seeds and ginger," says Hong. "But the problem is that consuming these foods directly doesn't do any good for the body and they're hard for the body to absorb, so we need a delivery system that increases their bioavailability." He is particularly interested in preventing diabetes and obesity, and asks how we can use foods like ice cream and chocolate without causing obesity.

In response, replacing fat with fiber and carbohydrates could help solve this problem, and for diabetes, better alternatives to sugar need to be found.

If Hong or others can succeed in creating tasty foods that contain fat substitutes, or by using nanoparticles to prevent the body from absorbing and storing fat and calories, they will have achieved the ultimate goal in food. "Many companies are researching foods that fill you up but don't affect your weight," says Hong, "But details are not yet available due to confidentiality agreements."

Some research into Nano

Increasing shelf life and improving the economic balance of food packaging

Packaging films and PET bottles with synthetic compounds are available on the Swiss market. Nanomaterials improve the carrier properties of gases, water vapor and flavors, as well as improve thermal and mechanical properties. Nanotechnology in PET bottles helps to balance CO_2. In the production, transportation and recycling of nano-packaging bottles, less than a third of the greenhouse gas emissions are produced compared to aluminum and about 60% less than conventional glass bottles.

Do nanoparticles in the packaging enter the food?

Although nanoparticles can pass through the packaging into the food, this primarily depends on how the nano-layers are applied and this transfer cannot be prevented. In this case, there is still no evidence that the product is harmless. This packaging can also be used as packaging with so-called antimicrobial action. A coating layer with antimicrobial nano-silver particles causes food to spoil more slowly.

Nano-index additives in nano-food

Today, there are only a few food products in Switzerland that are fortified with nano-specific additives. These include anti-caking agents that stop the thickening process of the condiments. This includes silicic acid (silicon oxide or (E551)) which decomposes into powders containing nano-sized particles. Nano-synthetic compounds are also used for encapsulation. For example, they are combined with carotenoids or vitamins to make them more soluble in water, more stable or better absorbed by ions. These additives also improve safety.

Improving food packaging

US researchers have found a way to keep food fresher or for a longer period of time, using less packaging material. Nanotechnology is a method of controlling the properties of materials that produce unique materials. Using a nano-structure, gas and water vapor can be designed to preserve fruits, vegetables, and beverages and other foods. Using nanoparticles, bottles and packaging can be made lighter and stronger, with better thermal efficiency and less gas absorption. These properties can extend the shelf life of products. Increase gas permeability in nano-packaging

The researchers found that polyethylene oxide (PEO) crystallizes when confined as a nano-layer and, as a single layer, resembles an impermeable single crystal, reducing the gas permeability of the polymer by 100 times. The researchers also noted that using a co-extrusion process, a new layer is grown that is melted into the polymer and combined. The growth of the layers to four or more layers creates a new structure that is progressively thinner and therefore more compact in the material.

Nano-packaging in contact with food

One of the commercial applications of nanotechnology in the food sector is packaging. Between 400 and 500 nano-packaging products are estimated to be in commercial use. Nanotechnology predicts that these products will account for 25% of all food packaging in the next decade. A major goal in nano-packaging is to increase shelf life by improving barrier performance in reducing gas, moisture exchange and UV radiation.

Over 90% of nano-packaging is based on nanocomposites, which improve the carrier properties in plastic wrappers for food and plastic bottles for soft drinks and fruit juices. Nano-packaging can also have antimicrobial, antioxidant and shelf-life extension properties, etc. In the future, the world of nano coating for materials such as cheese, meat and fruit, nano-packaging for chemical release, nano-based antimicrobial packaging and nano-sensors and nano-biodegradable packaging, nano-stick nano-layers for sauce bottles should be discussed.

Can nano diamonds be used?

This is an interesting area that has a lot of scientific interest and research, but its use in the real world is not worthwhile unless the current price and cost can be reduced by about 1.50. One of the problems with the food industry is that they don't like to spend money on the best quality packaging possible, so when new packaging comes along that costs 50 times more than the regular plastic trays they currently use, they don't want to do it, no matter how much better the packaging is.

There are claims that nano diamonds in food packaging can provide increased oxygen-barrier and antimicrobial properties to the packaging and while there may be more reason to use them in pharmaceutical packaging (although this is still unlikely), for food packaging this is unlikely to happen now or in the near future, mainly due to the price that manufacturers have to pay for their packaging.

Regulatory issues

In general, food manufacturers are required to ensure that their food products do not cause harm to the consumer from microbial overgrowth. While nanomaterials can offer a way to prevent this, they also need to be tested to ensure that toxic levels of nanoparticles do not migrate from the packaging into the food.

While each nanomaterial is assessed individually, the general consensus is that for a cubic liter of packaging containing 1 kg of food, the upper limit of migration is 60 mg per kg of food. There are many factors including: The type of nanomaterial used and the food, with studies showing that acidic foods and heating in the microwave (more so than on the stove) cause nanomaterials to migrate into food.

More studies are currently underway in this area, as some metal nanoparticles show migration values above the permitted threshold, while others are below the permitted limit, so there is a need for clearer contamination and nanoparticle migration results before nanomaterials can be widely used in food packaging.

New technologies in food packaging

Modern technologies in food packaging offer various methods for packaging. Below we mention some of these methods:

1. Active and intelligent packaging: This method can evaluate the conditions inside the package using sensors located inside the package. If the conditions inside the package damage the quality of the food for any reason, the relevant intelligent system changes the packaging to a defined state. This method can act as an effective method in maintaining the quality and increasing the shelf life of food.

2. Nanotechnology packaging: This method has the ability to improve the textural properties and maintain the quality of food by using nanomaterials. For example, by using nanotechnology packaging, food can be protected against oxidation, moisture and destructive bacteria.

3. Packaging using composite coatings: This method has the ability to maintain the quality and increase the shelf life of food by using coatings made from a combination of different materials (such as polymers, metals, etc.). For example, composite coating packaging can be used for foods that need to be kept cool.

4. Packaging using electronic printing: This method uses electronic printing on the packaging to display information about the contents of the package. For example, using electronic printing, information about the expiration date, contents, etc. can be displayed on the packaging.

5. Packaging using laser: This method uses a laser to make precise cuts on the packaging. This method can be used as an effective method for packaging foods with special shapes.

6. Packaging using robots: This method uses robots to provide automatic packaging. For example, using robots, products can be packaged automatically.

New packaging for frozen foods

New packaging for frozen foods can include the following methods:

1. Reactive packaging: These packages include files in which air and moisture are introduced into the package and then the package is placed in a cool environment and transferred to the refrigerator. This method can help preserve the taste and texture of the food more, which is usually used for meat.

2. Pressure packaging: In this method, the food is placed in high-pressure packages, which reduces the volume and size of the package. This method can improve the shelf life of the food, because the pressure causes less air and moisture to enter the package.

3. Nanotechnology packaging: In this method, the food is packaged using nanoparticles that are added to them. These nanoparticles can reduce the oxidation process and the growth of bacteria in the product.

4. Cold-affected packaging: In this method, food is placed in packaging at very low temperatures (such as -80 degrees Celsius). This method can reduce the evaporation process and bacterial growth and increase the shelf life of food.

5. Packaging using various gases: In this method, food is placed in packaging with special gases. These gases can act as food preservatives and increase the shelf life of food.

Packaging materials with modern technology

New packaging using modern technology may be made of different materials. Below I will mention some of these materials:

1. Packaging using recyclable files: These packaging is made of different materials such as polyethylene and polypropylene and are recyclable. This method can be used as a more sustainable solution for food packaging.

2. Packaging using biological materials: These packaging is made of materials such as cellulose and casein and are naturally biodegradable. This method can be used as a more sustainable and environmentally friendly solution for food packaging.

3. Packaging using nanoparticles: These packaging is made of nanoparticles such as nano silica and nano carbonate and can act as food preservatives. This method can increase the shelf life of food.

4. 3D Printed Packaging: These packages are produced using 3D printing technology and can be designed in various shapes. This method can improve the visual appeal of the package.

5. Smart Sensor Packaging: These packages have sensors that can provide information about the condition of the food, such as temperature and humidity, to the user. This method can improve food safety.

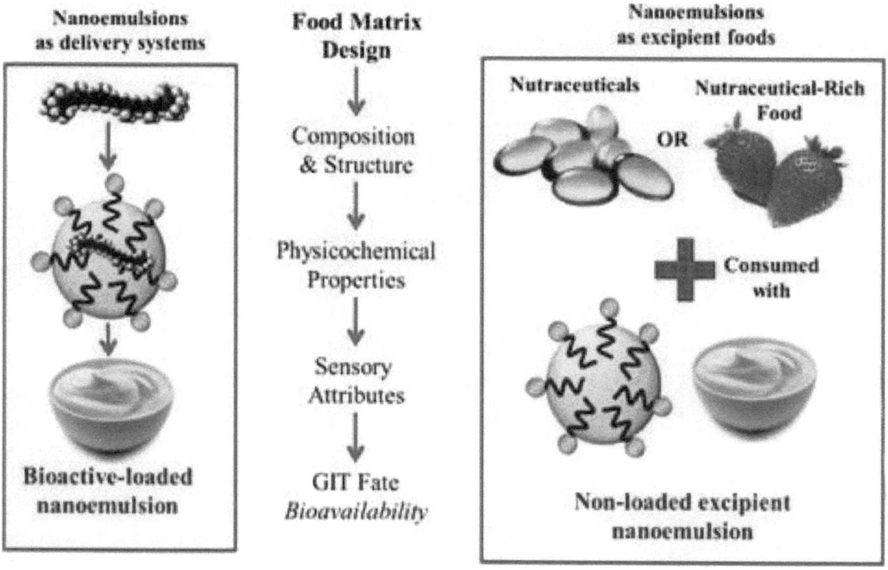

Figure 13. How Nanotechnology is used in the Food Industry

Chapter IV

A Review of Challenges Facing the Food Packaging Industry

The most problematic issue in food packaging is oxygen. Oxygen causes discoloration in food and spoilage of the fat in it, reducing the durability and life of food products. To solve such problems, the best solution is to use nanotechnology in food packaging. Nanotechnology offers several different ways to prevent food spoilage, including:

- For polymer food packaging, nano particles are placed in a zigzag pattern to prevent oxygen from entering the packaging. This maintains the freshness and durability of the food.
- To preserve lip products, nano uses tubes called nano filters to create an electric current in them, and it acts as an intelligent system in the grooves created by the oxidation of fats and passes through these pores. These nanotubes have the ability to separate molecules such as vitamins, proteins, lipids, and minerals based on their appearance and size.
- Coatings using cellulose fiber produced from renewable resources such as starch, cellulose and chitosan have the potential to be an alternative to non-recyclable materials such as plastic bags and non-renewable petroleum-based bags. In this type of packaging, food has become popular in recent years due to its environmental friendliness, cost-effectiveness, and protection of food products against spoilage and quality maintenance. This type of coating prevents the entry of harmful substances and factors, including oxygen, light, water vapor, microorganisms and other harmful substances. To increase the barrier and mechanical properties of cellulose packaging, cellulose micro fibrils are used for packaging. By using this type of coating in packaging, the penetration of air and oxygen into it is significantly reduced. One of the disadvantages of cellulose micro fibrils is water absorption, and in the presence of water vapor, several problems arise. To solve this problem, harmless and hydrophobic chemicals are used.
- Various methods have been tried to improve the coating of packaging, and coating with antibacterial paper is one of these methods that has passed the test well. This type of coating creates a surface with hydrophobic and absorbent

properties. Other types of coatings include nanocomposite coatings. Although this type of coating is very thin, it has high strength and is scratch-resistant.

By combining particles such as titanium dioxide or zinc oxide, we will be able to design coatings with very high resistance that are even impermeable to ultraviolet radiation. This type of nanocomposites is used to produce bottles because it has the ability to significantly prevent the release of beverage gas and is also resistant to oxygen penetration. Another application of packaging using nanocomposites is the packaging of food products such as meat, cheese, cereals, fruit juice, etc., which makes them durable for 18 months.

- Smart packaging using nano sensors: Nano sensors are often used to detect the presence of microbes or harmful chemicals. Their effectiveness is that since they are sensitive to toxic gases, they react to their presence, which causes them to change color and informs the buyer about the presence of fungi or toxic substances.
- Normally, labels containing petroleum materials that are non-renewable are used in product packaging. But today, with the use of nano technology, adhesives are used that contain starch and corn particles that are recyclable and do not harm the environment.
- Ice cream packaging is one of the most sensitive types of packaging. Its packaging must be in such a way that it reflects heat and sunlight so that the ice cream melts later.

Nano Features

Currently, most countries in the world have realized the distinctive features of nanotechnology to the point that they are making large investments in it and receiving a lot of benefits. The Food Standards Agency (FSA) has conducted a lot of research to achieve the ability to use nanotechnology in food, and specifically food packaging.

The British government has provided this company with a large budget for the development and continuation of advertising, and food delivery systems and methods for optimizing food appearance such as color, taste and consistency have been considered in this research.

Figure 14. Food Packaging is bad for the Environment

Given the increasing impact of nanotechnology on the food industry and the use of these products in the consumer market, the importance of the health of this category of food is being considered more.

The increasing need for this technology is further strengthening the acceptance of nanotechnology in sensory applications, and in this way, it is possible to be assured of the health of food. Exactly like a type of technology that alerts buyers and sellers that the expiration date of food is approaching. Antibacterial coatings and plastic bags that repel contamination have made a significant evolution in ensuring the health and safety of packaged foods. It is true that nanotechnology has received a lot of attention in the food industry and existing products in the market, but there are still many capabilities in this technology that can be extracted from nanotechnology.

Research conducted by the Institute of Food Science and Technology in the UK, in the field of nanotechnology, shows that more evolutionary research is still needed in this field before it can be used in food.

This research shows that current laws do not force companies to label items containing nanoparticles, so there is little chance that consumers will be able to know about their presence in food. Considering the size of these particles, along with the type and amount of ingredients, it is possible to determine their benefits and harms. According to research conducted by this English group, some major companies and prestigious universities have been accused of trying to monopolize new foods (through patenting). Because this practice causes problems for many innovative companies in developing countries.

There are no smart packaging and atmosphere-regulating packaging inside the packaging, and all products are packaged in ordinary polyethylene bags. On the other hand, equilibrium packaging inside the packaging is offered in international markets and exists unofficially in Iran.

Nanotechnology is one of the latest technologies that is used in many industries today and has been widely used in various fields. Using this technology in these packages prevents the contents from drying out and protects them from moisture, oxygen, and other hazardous substances. Nanomaterials have also been used to protect food from toxins and spoilage. With the knowledge of the capabilities of nanotechnology, it is hoped that it will be able to quickly change packaging systems and products and to preserve and preserve products more reliably using this technology.

Improved packaging

Nanoparticles used in improved packaging increase flexibility, increase gas barrier, and increase the thermal and moisture stability of the polymers formed from them. The use of nanotechnology in food packaging protects food from pathogens and harmful gases. Nano sensors are used to detect contamination and evaluate packaged materials, as well as protect food from ultraviolet radiation.

On the other hand, this technology improves food packaging and its capabilities and can be used to detect bacteria inside food packaging, which creates better taste and quality of food, increases safety and barrier properties against gas and moisture penetration, increases shelf life and prevents food spoilage.

Nano sensors

The main function of sensor-equipped packaging is to provide information about the environment inside the packaging, the condition and quality of the product, exchange information with external databases and provide information for subsequent decision-making. One type of sensor used for this purpose is a biosensor. Biosensors can be made by methods such as chemically changing and sensitizing packaging labels, embedding microchips (smart chips) or placing nano elements inside the label or the packaging itself.

Biosensors are an important choice in the agriculture and food industries, used to ensure food quality and safety with cost-effective and rapid methods, control products, processes, and detect gases in packaged foods for the integrity of packaged products. Various types of nanoparticles are used in the manufacture of biosensors, of which gold nanoparticles are one of them.

These materials provide advantages such as suitable size, low cost, fast response, accuracy and sensitivity. The purpose of using these materials in food packaging systems is to identify nanoparticles released in packaged food in a short period of time. Other applications include improving the sensitivity and ability to detect chemical compounds, bacteria, viruses and toxic substances. Nano sensors respond to environmental factors such as temperature, humidity, oxygen content, microbial agents and pathogenic microorganisms and provide useful information to producers and consumers about food storage conditions in terms of temperature, expiration date.

Nanocomposites

A composite is a bulk or mass that generally consists of at least two separate materials with complementary structures. In fact, polymer composites are mixtures of polymers and inorganic or organic fillers with specific shapes (fibrous, plate, spherical, and particulate).

Fillers are divided into three categories depending on how many of their dimensions are at the nanoscale:

- Spherical nanoparticles or so-called nano spheres (such as spherical silica nanoparticles), in which all three dimensions are at the nanoscale.
- Nanotubes are long structures in which two dimensions are at the nanoscale and the third dimension is at a larger scale.
- And when only one dimension of a constituent component is in the nanometer range, the composites are known as polymer nanocomposites. As a result, nanocomposites are a special class of composites in which at least one of the three dimensions (length, width or height) of the filler is in the nanometer range.

Nanocomposites are at the forefront of nanotechnology innovations related to the food packaging industry. A Japanese company was the first to put nanocomposites on its agenda for commercial use.

The company used montmorillonite clay prepared at the nanoscale to enhance the mechanical and thermal properties of nylon. Nanocomposites can be produced with thermoset or thermoplastic polymers (polyethylene, polypropylene, polyamides). These materials are highly reactive and are often used in food packaging. In general, the purpose of their use is to increase mechanical strength, reduce weight, increase heat resistance, better barrier against gases, UV radiation and moisture, better recyclability and greater product protection.

Bio-nanocomposites

In addition to having a nano-sized compound, bio-nanocomposites have biodegradable compounds and are converted into their micro-units in the environment by decomposing organisms. Among the biodegradable polymers, the following can be mentioned: Starch and its derivatives are important natural polymers that usually turn into an elastic material with low mechanical strength and poor protection against oxygen and moisture after compression, but their combination with nanoparticles causes the tensile strength to be developed and the water vapor permeability to be reduced.

Bio-polyesters are another important class that are composed of biological monomers including polylactic acid, polycaprolactone. The ability of these biopolymers as shells and molds around food has made them attractive materials in the industry. The only problem is their fragility and poor gas barrier properties; to overcome this problem, nano clay is used as a filler and support. The packaging must be able to preserve the encapsulated prebiotics, probiotics, and vitamins in the proper conditions, which these biomaterials have the necessary capabilities for.

Food materials and products are subject to various types of spoilage (chemical, physical and biological) and must be protected from environmental factors until the final consumption stage.

Coating and packaging play a major role in maintaining the qualitative and organoleptic characteristics of food during the storage, transportation and distribution process. In recent years, packaging has gone beyond its initial role as a food protector and is considered a key factor in increasing shelf life and safety, attracting customers and increasing competitiveness, raising awareness about the quality and characteristics of the product, marketing, branding and overall giving identity to the product.

Social and cultural changes that have taken place in recent decades, such as the increase in the urban population, increased purchasing power, changing lifestyles and a greater preference for ready-to-eat food products, changing the age structure of the population, the increasing importance of society to health, hygiene and food safety, and the globalization of food trade, which has led to an increase in the distance (spatial and temporal) between the production and consumption stages, are only some of the factors affecting the increasing position of packaging in society.

The food and beverage industry accounts for a total of 69% of the production of packaging industries worldwide, and this large volume of consumption clearly shows the importance of the close relationship between the food and packaging industries. The food packaging industry, like other sectors of the industry, must turn emerging threats into opportunities, gain a suitable competitive advantage, and create added value for its products through creativity and innovation and by providing appropriate responses to the increasing changes in consumer market behavior and expectations. The major challenges facing the packaging industry in the food sector are as follows:

Changing consumer lifestyles

The increase in life expectancy and the decrease in birth rate are among the most obvious changes in society, which have led to a change in the age composition, in other words, the gradual aging of society.

The age group of people over 60 years old, due to physical limitations, have different demands from other age groups, which include things such as clarity of information on the package, ease of opening and removing the product from the package, the ability to reclose it, or the availability of the product in different sizes. Other growing changes include the reduction in family size and the formation of the phenomenon of single-person households, the increase in simultaneous employment of parents, the increase in the level of welfare, the increase in the tendency to consume ready-to-eat and frozen food products, which have paved the way for the formation of new expectations from the packaging industry, and include things such as small and single-person packages, ease of use on the go (on-to-go), and the ability to heat in the microwave.

Development and application of new technologies in the food industry

The development and widespread use of new technologies in the food industry, such as irradiation, high-pressure processes, pulsed electric fields, ultraviolet radiation, microwave heating and antimicrobial packaging, requires the packaging industry to design and produce coatings and packaging suitable for these technologies. For example, coatings that can be used in irradiation processes must have polymers that are resistant to decomposition by the energies resulting from irradiation, or in ultraviolet radiation processes, the package must have a suitable transmission coefficient for such rays, or the voids must be able to withstand the pressure and temperatures used in the sterilization process.

Environmental threats

The food industry is one of the major consumers of packaging and packaging and plays a significant role in environmental pollution, and for this reason, environmental compatibility and packaging dynamics have received more attention than ever before. Dynamic packaging must meet the functional and economic needs of the present without compromising the ability of future generations to meet their needs, which includes compliance with issues such as reducing volume, weight, and packaging layers (compact packaging), increasing energy efficiency, using renewable energy, using biodegradable compounds, recycled compounds (with safety considerations), reusable packaging, and not using compounds toxic to humans or the environment.

Safety and Biosecurity

The primary expected feature of packaging is to protect food from external contaminants and maintain the safety and health of food throughout the production cycle to final consumption. In addition to external contaminants, the decomposition of coating components and their migration into the food are other important issues in food packaging.

Other packaging characteristics that affect the quality characteristics of food and should be considered in the design and application of coatings include the speed and rate of moisture and oxygen exchange with the surrounding environment, the removal or absorption of aroma compounds that occur between the food and the surrounding environment, between the food and the coating, or between the components of the food itself.

In designing a suitable packaging, sufficient information should be used regarding the migration of food and coating components, the absorption or removal of volatile and moisture compounds, the permeability of the coating to gas exchange, and the kinetics of oxidation reactions and other destructive reactions.

Biosecurity and combating the possibility of tampering with packaged food are other new topics in food packaging. The use of package integrity indicators, product freshness or food spoilage, the possibility of using antimicrobial compounds, antioxidants, oxygen absorbers, ethylene, carbon dioxide, alcohol, are used in active and smart packaging.

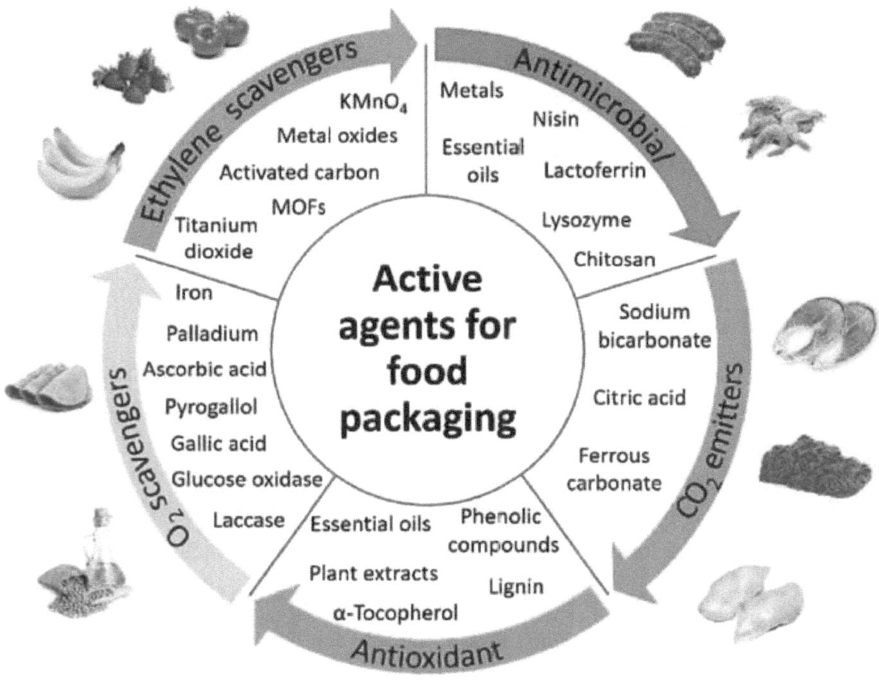

Figure 15. Food Packaging Materials with Special Reference to Biopolymers-Properties and Applications

Regulations and legal restrictions

The increasing strictness of national laws and regulations on the one hand and the globalization of food trade on the other hand and the requirement to comply with national and transnational standards have presented the packaging industry with new challenges.

Prohibition or restrictions on some materials and coatings make it necessary to replace and select suitable compounds with lower toxicity, volatility and migration under different environmental and processing conditions.

Economical

Economical is defined by the cost/benefit ratio for consumers and profitability for manufacturers. Increasing the benefits of packaging by improving efficiency and performance and reducing costs by implementing measures such as reducing the cost of raw materials, using high-performance equipment, and using compact packaging, reduces the volume and weight of the package, and subsequently reduces storage and distribution costs.

The increasing awareness of the consumer community about food safety and hygiene and the increasing level of competition in the food market have improved the position and importance of packaging in food. Packaging has gone beyond its primary function as a protector of food against physical, chemical, and biological stresses and has become an identifying factor for food, so that in most cases, a food product is recognized by its specific type of coating and packaging.

The research and development departments of the packaging industry must be able to respond to the expanding needs by employing expertise related to the food industry, microbiology, industrial design, graphics, marketing, and environmental engineering.

The first international Iranian packaging conference, which is being held by the Iranian Packaging Science and Technology Association and the Iranian Scientific and Industrial Research Organization, with the presence of domestic and foreign experts from universities and the packaging industry, will examine the challenges facing the packaging industry (including in the food industry sector) and present strategies.

Changing consumer lifestyles

Increasing life expectancy and decreasing birth rates are among the most obvious changes in society, which have led to a change in the age composition, in other words, the gradual aging of society.

The age group of people over 60 years old, due to physical limitations, has different demands from other age groups, which include things such as clarity of information on the package, ease of opening and removing the product from the package, reseal ability, or availability of the product in different sizes.

Other growing changes include the reduction in family size and the emergence of single-person households, increased simultaneous employment of parents, increased welfare levels, and an increased tendency to consume ready-to-eat and frozen food products, which have paved the way for the formation of new expectations from the packaging industry, including small, single-serving packages, ease of use on the go, and microwave heating capabilities.

Development and application of new technologies in the food industry

The development and widespread use of new technologies in the food industry, such as irradiation, high-pressure processes, pulsed electric fields, ultraviolet radiation, microwave heating, and antimicrobial packaging, require the packaging industry to design and produce coatings and packaging appropriate to these technologies. For example, coatings that can be used in irradiation processes must have polymers that are resistant to degradation by the energies generated by irradiation, or in ultraviolet radiation processes, the package must have an appropriate transmittance for such rays, or the voids must be able to withstand the pressure and temperatures used in the sterilization process.

Environmental threats

The food industry is one of the major consumers of packaging and packaging and plays a significant role in environmental pollution, and for this reason, environmental compatibility and packaging dynamics have received more attention than ever before. Dynamic packaging must meet the functional and economic needs of the present without compromising the ability of future generations to meet their needs, which includes compliance with issues such as reducing volume, weight, and packaging layers (compact packaging), increasing energy efficiency, using renewable energy, using biodegradable compounds, recycled compounds (with safety considerations), reusable packaging, and not using compounds toxic to humans or the environment.

Safety and Biosecurity

The primary expected feature of packaging is to protect food from external contaminants and maintain the safety and health of food throughout the production cycle to final consumption. In addition to external contaminants, the decomposition of coating components and their migration into the food are other important issues in food packaging. Other packaging characteristics that affect the quality characteristics of food and should be considered in the design and application of coatings include the speed and rate of moisture and oxygen exchange with the surrounding environment, the removal or absorption of aroma compounds that occur between the food and the surrounding environment, between the food and the coating, or between the components of the food itself. In designing a suitable packaging, sufficient information should be used regarding the migration of food and coating compounds, the absorption or removal of volatile compounds and moisture, the permeability of the coating to gas exchange, and the kinetics of oxidation reactions and other destructive reactions. Biosecurity and dealing with the possibility of tampering with packaged food are other modern issues in food packaging. The use of package integrity indicators, product freshness or food spoilage, the possibility of using antimicrobial compounds, antioxidants, oxygen absorbers, ethylene, carbon dioxide, alcohol, are used in active and intelligent packaging.

Legal regulations and restrictions

Increasingly stringent national laws and regulations on the one hand, and the globalization of food trade on the other, and the requirement to comply with national and transnational standards, have presented the packaging industry with new challenges. Banning or restricting some materials and coatings requires the replacement and selection of suitable compounds with lower toxicity, volatility, and migration under different environmental and processing conditions.

Economical

Economical is defined by the cost/benefit ratio for consumers and profitability for manufacturers. Increasing the benefits of packaging by improving efficiency and performance and reducing costs by implementing measures such as reducing the cost of raw materials, using high-performance equipment, and using compact packaging, reduces the volume and weight of the package, followed by storage and distribution costs. Increasing the level of awareness of the consumer community about food safety and hygiene and increasing the level of competition in the food market have improved the position and importance of packaging in food.

Packaging has gone beyond its primary function of protecting food from physical, chemical, and biological stresses and has become an identifying factor for food, such that in most cases, a food product is recognized by its specific type of coating and packaging. The research and development departments of the packaging industry must be able to respond to the expanding needs by employing expertise related to the food industry, microbiology, industrial design, graphics, marketing, and environmental engineering.

At the first international Iranian packaging conference, which is being held by the Iranian Packaging Science and Technology Association and the Iranian Scientific and Industrial Research Organization, and with the presence of domestic and foreign experts from universities and the packaging industry, the challenges facing the packaging industry (including in the food industry sector) will be examined and strategies will be presented.

Examining some of the most important challenges of the packaging industry

- ✓ **The challenge of material waste**

One of the biggest challenges in the packaging industry is material waste. This does not only mean packaging material waste, but also product waste. Optimal packaging material management methods, improving packaging technologies, and using high-efficiency packaging materials can help reduce material waste.

- ✓ **The challenge of maintaining product quality**

Due to the type and characteristics of products, maintaining quality over time and under different transportation conditions is one of the challenges of the packaging industry. This requires new packaging technologies such as nano-films, moisture-absorbing materials, and thermal management methods.

- ✓ **The challenge of environmental sustainability**

The packaging industry can have significant impacts on the environment, with high waste and energy consumption, including impacts and solutions such as the use of recyclable packaging materials, optimizing packaging design, and reducing energy consumption in production processes can help with environmental sustainability and solving the challenges of the packaging industry.

- ✓ **The Challenge of International Standards**

With the development of global markets, compliance with international standards in packaging design and production has become more important. Compliance with safety, health and quality standards can help companies attract global markets.

- ✓ **The Challenge of Changes in Consumer Demand and Tastes**

With changes in consumer tastes and needs, the packaging industry is also forced to respond to changes. This can bring challenges such as producing lightweight packaging suitable for single-use and using new packaging design and printing methods.

- ✓ **The Challenge of Packaging Security**

Packaging security is a critical issue in the packaging industry, especially in sectors where food or pharmaceutical products are produced. Countering counterfeiting and fraud, improving packaging systems with security features such as chip labels, and upgrading quality control systems can help increase packaging security.

✓ **Challenge of Innovation and Technology Development**

With the advancement of technology, new opportunities for improvement and innovation have been created in the packaging industry. Among these opportunities are the use of 3D printing, intelligent packaging image planning, and the development of biodegradable packaging materials, which improve quality, efficiency, and address challenges in the packaging industry.

✓ **Challenge of Changing Laws and Regulations**

Changes in laws and regulations related to packaging create challenges for manufacturers and distributors. For example, new regulations on the packaging of food products or pharmaceuticals may require changes in production and packaging processes, which can increase costs and time. To manage the challenges of the packaging industry, companies need accurate and up-to-date information on packaging-related laws and standards and must work on solutions that are in line with these changes.

✓ **The challenge of production and packaging costs**

Due to intense competition in the markets, reducing production and packaging costs is a significant challenge for companies. This is especially important for small and medium-sized businesses that are looking to optimize their processes to increase their competitiveness. Using new technologies, optimizing the supply chain and reducing waste can help reduce costs.

✓ **The challenge of coordination in the supply chain**

Coordination in the supply chain from production to distribution is one of the challenges of the packaging industry. Any shortcomings in this coordination can lead to delays, additional costs, and even reduced product quality. In order to optimally manage the supply chain, the use of inventory management systems, artificial intelligence, and information sharing with partners in the supply chain can be useful.

- ✓ **The challenge of changes in consumer behavior**

Consumer behavior may change continuously, which can create new challenges for the packaging industry. For example, the increase in the tendency to buy through online stores and changes in social consumption patterns can lead to the need for new packaging that is tailored to these behaviors. In this process, market analysis and forecasting of consumer trends can help companies to best align with consumer needs.

- ✓ **The challenge of communication and branding**

Effective communication and appropriate branding are among the key factors in the success of a packaged product. Creating meaningful connections with consumers, conveying brand values through packaging, and drawing consumers' attention to products can help grow sales and increase brand awareness.

- ✓ **The challenge of using new technologies**

With the growth of new technologies such as artificial intelligence, the Internet of Things, and block chain, the packaging industry is also facing new challenges. These technologies can be effective in improving performance, increasing productivity, and creating new opportunities for the packaging industry, but at the same time they require continuous investment and training.

- ✓ **The challenge of adapting to sustainability trends**

Due to the increasing sensitivity to environmental issues, the packaging industry needs to adapt to environmental sustainability standards such as reducing material consumption, recycling, and using renewable materials. This challenge requires changes in production and packaging methods, as well as cooperation with other members of the supply chain.

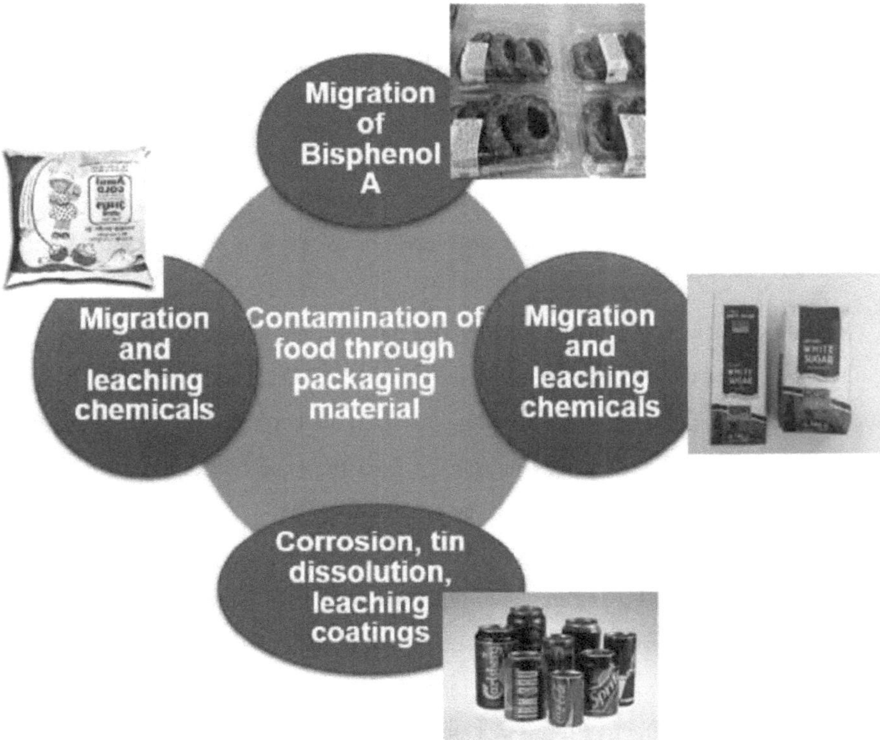

Figure 16. Food Packaging Materials with Special Reference to Biopolymers-Properties and Applications

Food Packaging Challenges

Food packaging is one of the key steps in maintaining the quality, safety, and attractiveness of products, but it comes with several challenges. One of the most important challenges is maintaining the freshness and shelf life of food. Packaging must be able to prevent the penetration of air, moisture, and light to prevent premature spoilage of the material. This is especially critical for fresh food such as fruits, vegetables, and meats. Although plastic packaging is popular due to its affordable price and waterproof properties, it has a detrimental environmental impact. Increasing consumer awareness of the negative effects of plastics has led to an increase in demand for biodegradable or recyclable packaging. However, these types of packaging are often more expensive and may be less effective at preserving food quality.

Food safety is also a major concern. Improper packaging can contaminate food with microbes or chemicals. In addition, packaging materials must not chemically react with foods and not introduce hazardous compounds such as BPA into the food.

Another challenge is visual appeal and marketing. In addition to maintaining safety and quality, packaging must be attractive to the consumer and provide clear and accurate product information. This balance between attractiveness and efficiency is one of the most important factors for success in today's competitive market. Reducing costs and managing resources are also important issues for manufacturers. They must seek solutions that are not only of appropriate quality, but also economical and environmentally friendly in order to remain competitive in the global market.

Packaging Industry Challenges

The packaging industry faces several challenges that can have profound impacts on the production, distribution, and consumption of products. Some of these challenges include:

1. Environmental protection and sustainability: One of the biggest challenges for the packaging industry is the widespread use of non-biodegradable materials such as plastics. Increasing environmental concerns have led consumers and governments to put more pressure on manufacturers to produce environmentally friendly packaging. Using recyclable and biodegradable materials as alternatives to plastics, although beneficial, is usually more expensive and may not have the required durability and performance.

2. Safety and hygiene: Packaging must protect products from contaminants, microbes, and external factors. However, in some cases, packaging materials themselves may be a source of risk. For example, there are concerns about the release of chemicals from plastics into food products. Also, the safety of food packaging must be in line with international standards, which can sometimes be costly to comply with.

3. Costs and economic pressure: Producing packaging that is both functional and aesthetically pleasing requires high costs in research, development, and raw materials. Rising costs of raw materials such as paper, plastic, and metals can put pressure on companies' profitability. Also, changes in government laws and regulations related to packaging can impose additional costs on manufacturers.

4. Technology and Innovation: The packaging industry needs continuous innovation to improve efficiency and reduce waste. The use of new technologies such as smart packaging (which can respond to changes in temperature and environmental conditions) and active packaging (which extends the life of the product) can help to upgrade the industry, but at the same time it requires large investments.

5. Waste management and recycling: The increase in packaging waste is another important problem. Many countries are facing waste and recycling crises, and the packaging industry must look for solutions to reduce waste generation and increase recycling rates. These challenges show that the packaging industry needs creative and sustainable approaches to be able to adapt to consumer needs and environmental and economic constraints.

Futures scenarios, the study of the future in the present?

One of the tools used for futures studies is scenario writing. In order to take advantage of upcoming opportunities and make accurate predictions in today's world, which is full of ambiguity and uncertainty and is changing rapidly, scenarios are valid as an efficient and useful tool to prepare for uncertain futures, change mental models, test decisions, and improve performance in a dynamic environment.

In our country, due to the existence of numerous and diverse food industry companies, the packaging industry is considered a driver and enabler for the country's economy and is a driving force for the development of domestic production, increasing exports, improving the economic situation of the society, creating jobs, and increasing national income. In the past, packaging was used only to protect the product, but the modern and contemporary packaging industry can be considered one of the achievements of the industrial revolution.

In fact, packaging is a type of technology and a coordinated set of materials, energy, value, and thought that has a protective structure and, on the other hand, plays an informational role for the product. Today, packaging has a tremendous impact on the customer's perception of the product.

Studies show that 70% of product purchase decisions are made at the point of purchase and packaging plays an important role in consumer purchasing behavior. Packaging can have significant effects on the hidden dimensions of the product. For example, it can be said that in today's stores, the seller does not have much influence on the purchasing process and customers rely more on their own information or the information they get from the packaging when making a purchase.

In addition, today, packaging has developed from a silent seller to a brand creator, and effective and convincing packaging is considered the best way to present the product to the consumer. It can be said that packaging, as one of the dynamic industries, plays a prominent role in the development of economic sectors. Good packaging goes beyond the seller and is a recognized banner and symbol of value, which is why packaging design is considered a vital tool for customers in product selection and an opportunity for companies to increase commercial value.

Futures research is a systematic approach and includes a set of efforts that, by using the analysis of sources, factors and patterns of change or stability, visualize potential futures and plan for them, and has a huge impact on industrial competitiveness, wealth creation and quality of life.

In fact, it can be said that futures studies move from external forces that influence the future (astrology and prediction) to the structure of historical patterns of change, the emergence and fall of nations and systems, and human factors to the study and creation of desirable futures. The history of futures studies in Iran goes back to the country's five-year plans and the country's development vision document on the horizon, 2025, Iran's first strategic thinking and futures research document.

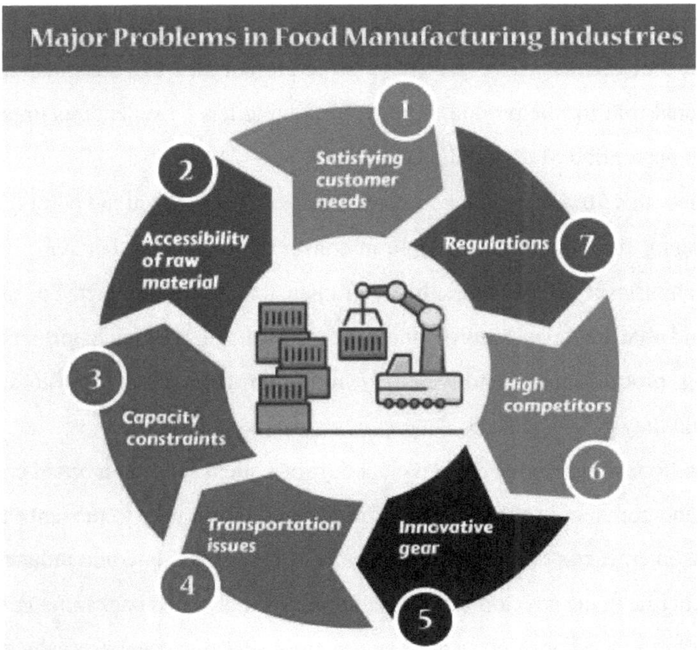

Figure 17. What are the challenges faced in food manufacturing industry

The impact of innovation on the food industry

Despite the obstacles to innovation in the food and beverage industry, innovation has had a significant impact on the growth and development of the industry. New technologies, such as biotechnology and information technology, have enabled manufacturers to introduce new and attractive products to the market and improve the customer experience.

1. Improving the quality and taste of products

Innovation in raw materials and production processes has improved the quality and taste of food products. For example, the use of new technologies such as advanced distillation or fermentation has helped produce better quality beverages and distinctive flavors.

2. Increasing efficiency and productivity

Innovation in production processes has increased productivity and reduced costs. The use of robots and automation in production and packaging processes has reduced production time and increased product quality.

Innovation Trends in Food and Beverage

Innovation in the food and beverage industry is not only driven by changing consumer needs, but also by the opportunities presented by emerging technologies and trends. Several important trends are emerging in this area that will broadly impact the future of the food and beverage industry:

1. Use of innovative raw materials

With increasing attention to health and sustainability, food and beverage manufacturers are looking to use raw materials that have a lower environmental impact in addition to high nutritional value. For example, the use of plant-based proteins as an alternative to animal proteins has become a prominent trend in this industry. Companies that are able to develop plant-based products with similar taste and quality to animal products can succeed in this competitive market.

2. Innovation in Packaging

Innovative packaging is one of the most important areas of innovation in the food and beverage industry. The use of renewable and biodegradable materials has increased in recent years, and smart packaging, capable of providing accurate information about expiration dates and product quality, has become one of the most exciting innovation trends in the industry.

3. Customization and personalized experience

One of the most important changes in consumer behavior is their desire to customize food products based on their specific needs. Food companies are using new technologies to provide customized products.

For example, gluten-free, lactose-free or low-sugar products can appeal to consumers with specific dietary needs. Also, the use of applications and digital platforms allows customers to easily order personalized food and drinks.

Barriers to Innovation in Food and Beverages

1. Technological Challenges

Despite significant advances in food industry technologies, many companies still face technological challenges in production and innovation processes. Applying new technologies to create new products or improve production methods requires advanced infrastructure and expert teams, which require high investment. Smaller companies often do not have the financial resources necessary to develop advanced technologies. For example, technologies such as artificial intelligence, the Internet of Things (IoT) and 3D printing have great potential to change the way food is produced and distributed, but implementing these technologies is a major challenge for many smaller manufacturers. Also, the lack of unified standards for using these technologies can create further barriers to innovation in this industry.

2. Changing consumer needs

Today's consumers care more than ever about the quality, authenticity and sustainability of food products. This change in tastes and expectations is forcing companies to update their products and adapt to new needs. But adapting quickly to these changes is a major challenge in itself. For example, the desire to consume organic and preservative-free products is increasing day by day. Companies that have previously relied on mass production and the use of chemicals and preservatives may face difficulties in adapting to these new needs. In addition, increasing awareness of environmental issues is also putting more pressure on companies to offer more sustainable products. Such changes require investment in research and development, and companies must respond to these needs while maintaining profitability.

3. Legal and regulatory constraints

One of the biggest obstacles to innovation in the food and beverage industry is strict laws and regulations. Governments impose specific and detailed laws and standards to protect public health and ensure the safety of food products. Although these regulations are necessary to protect consumers, they can significantly slow down the process of innovation.

✓ **Food safety standards**

Compliance with food safety standards is one of the main legal requirements in this industry. These standards, set by regulatory agencies, must be met at all stages of production and distribution. From raw material production to final packaging, companies must ensure that their products comply with safety regulations. This can slow innovation and increase costs.

✓ **International regulatory differences**

Manufacturers looking to enter global markets must deal with regulatory differences in different countries. For example, EU regulations can differ significantly from those in the US or Asia. Companies must invest significant resources in researching and complying with each country's regulations, which can be a major barrier to innovation.

✓ **Approvals required for new products**

New food products often require formal approvals from regulatory agencies before they can be marketed. These approvals are time-consuming and costly, and can delay the process of bringing new products to market. Innovative companies often face numerous challenges in this area.

Sources introduce different factors as the causes of success and challenges facing new product development. Among the factors effective in success are leadership in holding the vision and driving the project, as well as team cooperation.

Lack of proper management, lack of knowledge management in project management, lack of appropriate models in the project, lack of training, especially human resource training, and lack of sufficient knowledge of the customer and market are among the common challenges facing product development, especially food products. The type of product and food product, the cultural and economic atmosphere of the producing country, population composition, climatic conditions, available resources and infrastructure can be effective in the emergence and severity of each of the above challenges.

In most developed countries, the role of agriculture and related industries goes beyond providing the required food and also includes providing political, social and economic priorities. In this regard, the dairy industry, as the most important link in the food chain and the most advanced transformation and complementary industry, is receiving special attention.

This industry has occupied the first place in the consumption market and added value among transformation and complementary industries. This indicates the profitability and importance of policymaking in this area. On the other hand, dairy products with inherent nutritional properties have the ability to improve quality and have a greater impact on health by changing animal nutrition or formulation.

Therefore, it seems that considering the role of these products in the current and future food basket of people, recognizing the challenges of innovation in this industry will help provide better customer service, product diversification, quality improvement, and maintaining sustainability in the economic system and market. So far, limited studies have been published in this regard that provide a comprehensive, effective and practical assessment of all effective components in the production chain (farm to home).

A wide range of topics and obstacles have been introduced in the sources. These divisions differ according to the field of study (marketing and sales, product design, technology), with most of these studies focusing on the final stage, i.e. introducing the product to the customer.

Therefore, answering the question of what obstacles exist in the dairy industry from the time of idea generation until the new product reaches the consumer, or what solutions have been proposed for its comprehensive management, can be useful and effective.

Methodology

This review was conducted systematically based on library studies, scientific articles published in reputable databases including Scopus, Elsevier, Google Scholar, SID, news agencies and reputable domestic (including the Civilica database) and foreign conferences on key challenges.

Given the limited information focused on the food industry, especially the dairy industry, materials were selected from relevant Persian sources searched from 2011 onwards and for foreign sources from 2000 onwards. Challenges facing innovation, food industry, dairy industry, project management, design and packaging, marketing and sales of food products, new methods of process management in industries, especially the food industry, were considered among the keywords used to search the sources.

Then, out of 60 reviewed article and news titles, 43 titles were used for practical evaluation based on content relevance and searched headings. After summarizing, the extracted materials were classified and organized thematically based on the stages of product production until the product reaches the consumer. For this purpose, the basic concepts were first extracted, then the current state of innovation in the world's industries, especially the food industry in the country, was evaluated, focusing on the dairy sector. According to the collected materials, the challenges facing innovation were divided into 10 groups and examined.

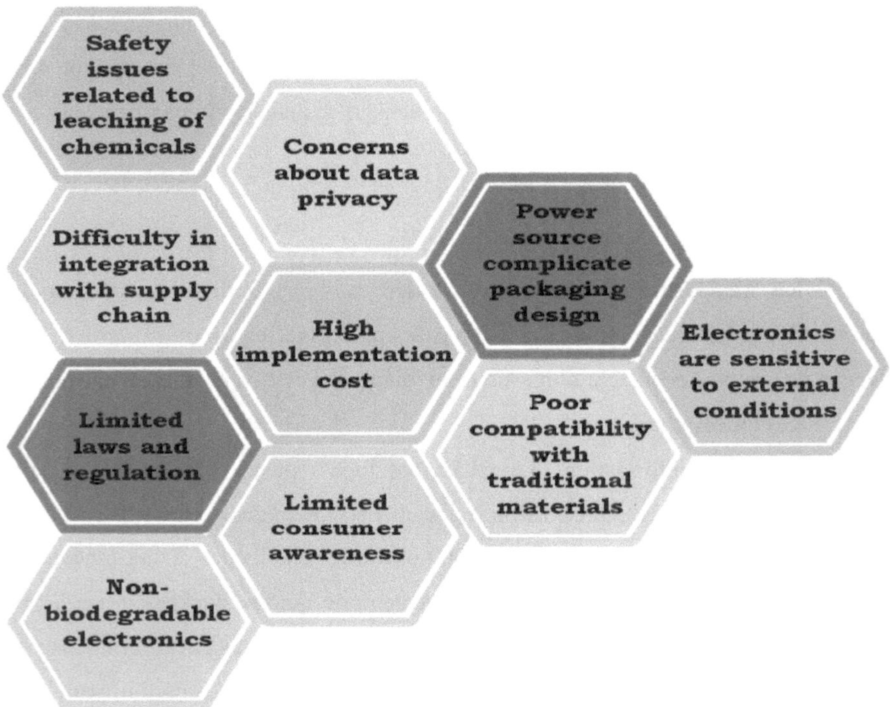

Figure 18. Recent technological advances in food packaging: sensors, automation, and application

The state of innovation in the world's and Iran's food industries, focusing on the dairy industry

Agricultural products create high added value due to the processing stages, so they are of interest. A study of economic indicators related to the food industry as an intermediary link between agriculture and consumers shows that the food industry has accounted for more than 12% of industrial workshops, 15% of employment, 12% of investment, less than 8% of added value, and 7% of industrial exports in the past 10 years. For example, in 2008 in Iran, among the subgroups of food industries, the largest share of added value with 22% belonged to the dairy industry.

Dairy industries around the world are facing various challenges such as animal health and welfare, food safety, sustainable management, climate change, and market liberalization, which require them to be constantly on the path of innovation. Accordingly, different countries have presented long-term and short-term plans to maintain sustainability and create an innovative environment.

For example, in the definition provided by the American Dairy Industry Innovation Center, the sustainability of this industry (dairy) is "Providing nutritious dairy products to consumers in a way that improves the industry, people, and the earth economically, socially, and environmentally for future generations (present and future)". In Iran, the dairy industry also has a high position and potential among the agricultural industries of the country and society.

The Iranian dairy products market is a competitive market that has many applicants for entry and investment. This shows the importance of proper management and policy-making regarding the use of the potential of this industry in the country. Studies show that innovation in the production of new dairy products will attract domestic and foreign markets, encourage livestock farmers and farmers to produce more, attract more human resources, provide milk with healthy quality, and produce products that suit tastes and diets.

On the other hand, identifying methods for utilizing dairy products in the formulation of other food and pharmaceutical products, including whey, will create greater prosperity in this industry.

Another form of innovation in the food industry, especially dairy, that we are witnessing in the world and in the country is providing products under the title of healthy or functional food products. In addition to nutritional value, these products will have therapeutic and health effects, such as improving lactose digestion, enhancing the immune system, improving the absorption of vitamins and calcium, inhibiting gastrointestinal infections, etc., including types of probiotic yogurt. Biotechnology knowledge has also created this opportunity for the production of functional foods, food additives, and micronutrients, and is being considered.

According to studies, health advertising also has a positive effect on the promotion and sale of food products. Many dairy products may face challenges in making health claims due to their inherent nutritional properties in terms of energy, saturated fat and sodium. However, new and innovative methods have been proposed to reduce these challenges, including changes in the diet of dairy cows.

For example, the effect of using extruded flaxseed on reducing saturated fats in milk and changing the fatty acid profile in favor of essential unsaturated fatty acids such as omega-3 and omega-6 can be mentioned. Therefore, considering this example, it seems that such health changes can be made without changing the taste and flavor of dairy products. What is clear is that organizations and industries today need structures to effectively move towards innovation.

One of the important elements in any structure is a proper understanding of the stages of innovation. A thorough innovation process ensures that all ideas received are evaluated and planned without any compromises. Identifying the types of problems and related challenges, providing broad solutions to these challenges, making careful decisions with ideas containing the program, sampling and implementing sample ideas, practical methods for applying ideas, and finally, careful evaluation of the innovation are mentioned as the steps of a comprehensive innovation program.

Getting the initial idea wrong and the idea evaluation process are common challenges associated with the innovation process. Companies and industries should start their innovation process clearly by identifying the critical issues and challenges ahead in order to get the best start. As a result, the energy of the personnel will be focused and purposefully spent on obtaining valuable ideas and providing solutions. The network of "Open innovation" groups is an advanced example of this, which expose their innovative needs to the public.

The challenge, in terms of personnel innovations and announcements by people outside the company or organization, will directly affect the innovation strategy. At the time of presenting an idea, the vision will be based on the success of the plan or that idea. A plan will be effective and successful when the future can be predicted and measured as much as possible.

At the beginning of an idea, cost and revenue projections remain uncertain because the innovation may be limited to a new business or an internal experience or even in the form of a new and emerging technology, as a result of which it has not yet been widely implemented and there is not enough information to estimate its cost. In this case, industries should consider more points in the real world for evaluation, identification and accelerate the learning and improvement cycle in the prototyping stage.

The appropriate structure of project definition, the perspective of public participation and interaction, the strategic approach and strategic formulation of new product development, product design based on hidden customer needs, management support from related industries and the government, careful study and research using reputable scientific sources in the world, having the necessary infrastructure (strong marketing and sales team, information technology, technology and expert human resources), stable and quality raw materials are generally considered effective factors in the commercialization process and innovation in the production of new food products.

Challenges facing innovation in the food industry

1- Designing and identifying appropriate resources for the project

Studies show that determining the appropriate time to provide the necessary resources is a major concern in managing the production of new and innovative products. The project office is responsible for determining the time required for each part of the project and implementing it in the initial planning of the project process after the idea is designed.

However, despite considering the cases, the figures and numbers in the design of the resource allocation process are not done in detail. For example, in many processes, the placement of people and the description of the defined services do not match the tasks, causing the figures and numbers to not match what happens in reality. In innovative projects, finding the right and sufficient resources is more difficult, especially in the production engineering and technology sectors, because generally in these two sectors, several projects are being implemented simultaneously, and resource limitations, especially financial resources, will be more apparent.

Delays in each stage, especially technology procurement, will dynamically and continuously affect all stages of new product production and will cause delays in the entire process. This will create a completely complex situation in a new and innovative project. Because most of the implementation and procurement process may be different from past processes and have little substitutability.

In such circumstances, due to the limited resources in the company, the focus will be on daily and routine production lines. The long study process and standardization of new and innovative products may result in a delay in the time of resource procurement of about 5 to 7 years, which will lead to the production of new products being removed from the company's priorities.

Therefore, project managers will have conflicts and disagreements with the existing organizational procedure for producing a new product. In this regard, it is necessary to connect the new product production process with the current process in the company. This will be difficult given that in the early stages of process design, the implementation path does not have sufficient transparency.

Providing resources for initial studies to examine the impact of a new and innovative project on the production process is another problem mentioned at this stage of new product production. The project team has the opportunity to make corrections based on the questions raised, but due to lack of resources and lack of proper forecasting, specific assessments to correct the process will face difficulties, which will also affect the quality and delivery time of the project. According to Mir's report (2014), almost all experts consider one of the most important challenges to the growth and development of the agricultural sector in Iran to be the lack of sufficient capital in this sector. According to another report, financing production projects is one of the challenges facing production in the food industry.

2- Time preparation and planning

Timeliness is the next challenge in managing new product development. This is particularly evident at the end of the project. Having the right equipment, relevant articles, produced samples, and quality assurance for the new product industrialization process at the right time is a must.

Most of the production and decision-making activities should be specified by the company's R&D department or project office before industrialization. Updating drawings and calculations and predicting possible deviations, as well as the tolerance level, should be re-evaluated because there is not enough time to experiment and learn from failure. Due to the wide scope of activities in some complex new product introduction projects, the possibility of delays in the process increases. This delay may have a less detrimental effect on product development, but it will have a greater effect on the start of the production process.

The next challenge in this area that will affect the project implementation time is the lack of coordination between suppliers and manufacturers. Delays in procurement or investment are one of the most effective factors that can pose a major challenge to the entire program related to project implementation, so coordination and assurance of supply and investment resources are of particular importance.

Bakeri (2015) also mentioned the inability to create a win-win relationship with stakeholders and the development of a tense relationship between dairy industries and raw milk production enterprises and livestock farms as one of the effective challenges in the country's dairy industry. The country's economic sanctions and the resulting lack of timely supply of some required parts and machinery are also among the effective confirming factors in this group.

3- Decision-making gates and the number of decision-makers

In general, most development projects include time for performance evaluation and process monitoring. These gates are of particular importance in project management. Budget evaluation and compliance with the schedule are among the measures taken at this stage.

On the other hand, many believe that these gates have become an administrative tool instead of supporting the project. However, what is clear is that if the project is implemented according to the plan, the gates can be a tool for reflection and decision-making to evaluate progress in terms of cost, quality and time. However, in most cases, fundamental changes, limitations and deviations from the project path prevent this from happening.

As a result, project managers will be forced to prepare the information needed to submit reports by omitting some project components. Therefore, the question arises as to how much decision-making gates will lead to added value in the project. Some projects will be stopped at these gates due to lack of agreement and decision-making at specific times, which will slow down the process of the new product production project. Some of these assessments are due to external regulations and requirements that the manufacturer is required to comply with in the product production process.

The instability of regulations and the lack of familiarity of some producers with these regulations have been introduced as challenges for the food industry in the country, which makes the process of reviewing and examining at the decision-making gates more difficult. Khodaei (2010) also pointed out in this regard the lack of awareness of the criteria for evaluating ideas (producing a new product or a new process), organizational struggles and politics, as well as the passing of suggestions and ideas as obstacles to innovation. In another study, the high volume of work in the project team and numerous external tests (managing requirements and standards) were mentioned as factors affecting the delay in introducing a new product in the food industry.

4- Old working methods

In feasibility studies of new product production projects, the level and degree of readiness for production (MRL: Manufacturing Readiness Level) is generally not considered; At this stage, the focus is on the level of readiness and degree of availability of technology.

In this regard, a limited number of project managers pointed out examples where in some projects, concepts were sufficiently mature from a technology and technology perspective, but were not sufficiently mature from a production perspective to implement or enter a new stage. The lack of structure of the project for the production of a new product can be seen in the interaction with technology, especially in the early and initial stages of production.

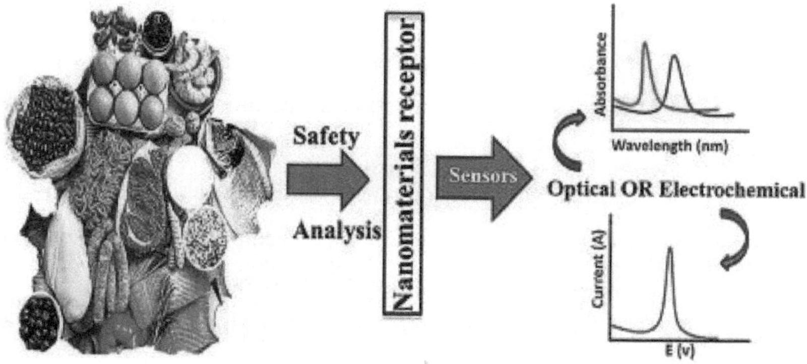

Figure 19. Food Packaging

For example, the most effective mental and program readiness in the implementation of a new production project is the use of non-specialists and people unfamiliar with the new technology, process or designed product, among these factors. This issue shows the role of education in updating the knowledge of employees and managers in order to change the traditional approach to a modern one in the production process, in order to better utilize the new technology.

Retraining of expert forces is considered one of the positive measures resulting from the relationship between industry and university, which unfortunately, this challenge and gap is palpable in most industries, including the dairy industry of the country. In his interview with Donyaye Eghtesad newspaper, Nabi also stated that the lack of trained, literate, and experienced workers in the food industry is a serious challenge.

Sometimes, the use of workers who lack the necessary knowledge in the field of work even leads to reduced efficiency, creating a polluted and inappropriate environment in food production.

Shafiee et al. (2012) stated in their research that today, the relationship between universities and industry is considered a mechanism for inventing technologies and producing new products, retraining specialized personnel, creating jobs, and global competition. In another report, the lack of communication and cooperation between research and development units of companies with customers and research centers is mentioned as one of the obstacles to innovation in the food industry, especially the dairy industry.

In another category regarding the effect of external communications on creativity and creating an innovative environment, cooperation with companies or organizations had a positive effect on innovation in the process and in the product. However, ties with academic and research institutions only had a positive effect on innovation in the product. This result shows that external factors and education can help create and grow an innovative environment, especially product innovation.

5- Poor communication and inappropriate time sharing

Establishing effective and good communication with production line managers and organizing the production line is considered a challenge for the production of many new products.

Often, due to the ambiguities that exist in the introduction of a new product or innovative process, it is not clear how to prioritize new projects. As a result, launching production for a new product sometimes creates conflicts in terms of providing resources and time with the daily production activity in the factory. Active experts have also acknowledged this conflict.

One of the reasons is the failure of the project management team to consult with all managers in estimating the time for using resources. The company generally deals with the common and daily production process, as a result, issues related to the new product or process are not prioritized in the production line.

The challenge will be much more complex in large and extensive projects that require the involvement of several departments in the factory to provide the required resources. According to experts, it is difficult to establish communication between all people related to the production chain due to time constraints.

For example, production engineers are active and involved in several parallel projects in addition to supervising the daily production process. Employees' lack of awareness of managers' expectations of how to participate and be creative, as well as the isolation of senior managers, will contribute to this atmosphere. Danesh Kohan and Bidgoli (2009) also reported that product development requires a multi-sectoral effort between all units involved in production.

6- Loss of opportunity to learn

What is clear is that the learning process from projects does not take place in companies. For this reason, the problems of one project are repeated in subsequent projects. Most learning takes place in informal project meetings, although the goal of this learning is to avoid similar mistakes in future projects in the long term. In general, it is recommended that at the beginning of each project, team members have a review of past knowledge and experiences, for example, they can study the report of the past six months (including on the budget in terms of time and cost). Another issue related to this section is the lack of effective follow-up of issues by experts.

Knowledge management is considered as a practical solution to overcome this challenge. Today, it is possible to access the experiences of others, thanks to the recording and organization of past and current achievements in the production process. Therefore, using standard methods and techniques to record and maintain experiences and establishing mechanisms to increase interactions between individuals in order to disseminate tacit knowledge and transfer experiences to others contributes greatly to increasing the success rate of products and, as a result, the growth of businesses in this industry. Knowledge activity is at the heart of product development processes and is of interest.

Many managers and business owners, especially in the food industry, are inclined to decide how to empower knowledge management or coordinate activities related to the production of their company's new product through knowledge management tools. In another study, lack of access to reliable, transparent information or excessive information has been mentioned as obstacles to the production of a new product or an innovative process.

7- Definition of economic activity

Defining the economic processes and activities is very important for managing a new product introduction project. In the early stages, there may be many discussions about the economic activity, for example, whether the company has the ability to produce a new product or make changes to the process, whether the required equipment is available or whether there is a need to provide and innovate in manufacturing, or whether there is a capacity in terms of physical space at the factory level.

In the first step, the project manager needs to spend more time to define the economy, study the documentation and obtain approval from other relevant managers, which can take a long time. The long process will lead to uncertainty and insecurity in the decision-making process.

8- Many projects with low competence

The competence of the project manager is another challenge in the new product production process. Most projects include a number of sub-projects, including new product production projects and investments. Sometimes, for small projects, it is easy to select a manager. As most managers in this role are new and inexperienced, in fact, new forces are selected as project managers. Therefore, in the first step, detailed training on project expectations is needed in terms of resources and outputs. Edako et al. (2018) also listed the existence of a large number of projects running simultaneously and the lack of a continuous management structure as factors that delay the production of food products.

9- Marketing, design and sales

Given the competitive environment of products in the food market, innovative activities are carried out at levels such as changing product compositions, using additives, offering variety in terms of taste, modifying packaging design, focusing on product positioning and creating market segmentation, marketing strategies and advertising. Marketing is the final stage in the chain of producing a new product and delivering it to the customer.

According to studies, this stage is considered one of the most important and challenging stages. As mentioned, in this stage, several topics are considered to introduce a quality and new product, including packaging, pricing, product introduction, branding, advertising, etc. Studies show that neglecting product marketing is one of the reasons for the decline in agricultural and livestock production. This will reduce the profitability of crops and, as a result, reduce the motivation of farmers and livestock breeders.

Donyaye Eghtesad, quoting Nabi, states that what encourages consumers to choose a product in a store, in addition to reputation and history, is the appearance of the product, that is, the packaging. According to this report, the packaging industry in Iran has made great progress in recent years, but due to a lack of investment in it, it is still far from meeting the necessary conditions to be present in global markets. In addition, containers produced domestically, which are mostly made of glass and metal cans, do not have the ability to be standardized, and this is one of the annoying problems that food manufacturers are dealing with.

On the other hand, paying attention to old elements in product design and providing formulations reminiscent of past memories are among these cases that are mentioned in the design and marketing process as retro marketing. An example of this in the dairy industry can be the design of new milk bottles with the same appearance as old bottles (the production of glass milk inspired by this knowledge by the Pegah Tehran Company for old customers) or the design of some products in containers with an appearance similar to milk jugs in some dairy industries around the world.

Digital marketing is also one of the current topics in this field. Today, marketing without the use of digital tools and channels is no longer as effective as before. Therefore, many dairy companies in the country have turned their attention to this issue. The need to pay attention to this marketing method is especially evident in critical situations (an example is the Corona pandemic and the need to reduce urban traffic for shopping). This marketing method makes sense by utilizing the various channels where today's customers spend most of their time. Advertising and sales are among the measures that are carried out through this method.

In order to find and maintain the presence of products in target markets, especially exports, attention to branding is of particular importance. The most important challenge in exporting and maintaining the presence of Iranian products in target markets is the lack of continuity in the presence of goods in this market, as the average presence of Iranian products is 2 to 2.5 years.

Brands are not only important in the target market, but also for the survival of products in domestic markets against the influx of foreign products, countries need to develop brands. In our country, there are diverse and unique products that are exported to different places without Iranian brands. For example, saffron exports to the world are often without packaging. Iran has export capabilities in many products, including fisheries and dairy products, but proper use of these opportunities and overcoming this challenge requires planning. Table 1 shows the key challenges facing product introduction in the food industry.

Table 1. Key challenges facing product introduction after production

Row	Marketing and sales challenges	Sources
1	Unknown product for the seller	http://keshavarzfazl.com
2	Lack of understanding of new product features by the buyer	Razvani and Rashidi Abandansari, 2013
3	Improper packaging	Ksenia, 2013 Nabi (2014)

4	Lack of attention to branding and use of unreliable brands	Karbasi and Aghasafari, 2016
5	Incorrect forecast of market demand	Barikani et al., 2007
6	Weak advertising	Cooper and Klienschmide,1993 Pishbahar et al., 2017
7	Product distribution channel	Cooper and Klienschmide,1993.
8	Incorrect forecast of competitor behavior	Cooper and Klienschmide,1993.
9	High cost price	Mortezaei, 2016 Bakeri, 2015

10- Environmental considerations

The results show that although environmental innovations play the least role among the innovations of the dairy industry, improving the sustainability of this industry is of great quality and importance, so that it has occupied the first two ranks of innovations in the dairy industry in recent years.

Recycling and controlling waste and wastewater, processing whey, and providing environmentally friendly packaging are among these. Some environmental threats arise from the activities of the dairy industry. Therefore, dairy companies and their affiliated units will be committed to innovation in the field of health and environmental protection. Dairy industries, like other industries, can have a negative impact on the environment by selling packaged products (especially non-biodegradable packaging). Since consumers have less opportunity to choose a product without waste or with less waste, this pollution occurs more quickly. Attention to environmental health sometimes encourages some customers to choose a larger product to produce less waste.

According to the Chamber of Commerce report (2017), Iran's dairy industry has faced a 37% decrease in dairy consumption over the past 7 years. In this regard, customer retention and expanding the range of consumers are among the concerns facing managers in this industry.

Obviously, recognizing the challenges facing production in new products and its comprehensive management will reduce investment risk. The modern world requires a new and up-to-date solution to produce a new and innovative product and to offer it to customers.

In the present study, according to the studies, the basic challenges in the entire new product production process were introduced and identified in ten categories. What is clear is that not paying attention to any of the challenges mentioned can have a great impact on the entire product production process from idea creation to the product reaching the consumer. For this purpose, comprehensive management of the entire process is necessary and essential.

The use of intelligent automated systems to manage production processes is of particular importance in modern conditions in order to make timely strategic decisions in product production. As the Russian Ministry of Agriculture is actively pursuing the implementation of digital economy programs in its country. This program is designed with the aim of transforming the agricultural sector and achieving technological success.

According to a study conducted in Russia, organizing management is the main condition and the main factor in increasing competitiveness in quality milk production. Bakeri (2015) also recommends creating a distribution and sales system with a credit electronic sales plan as one of the solutions related to the final stage of production in the country.

According to the findings, especially in the food industry, comprehensive operations management and support are needed to deal with perishable products, unpredictable changes in supply, food safety and creating stable conditions. One of the modern and efficient methods in the world that can be effective in monitoring all components of the food production and supply chain, including livestock products, is the use of Internet of Things (IoT) knowledge.

Virtualization enables supply chain agents to monitor, plan and optimize business processes remotely and without interruption via the Internet, as virtual objects, without being present on site. Therefore, it is suggested to create an integrated management system using modern methods to reduce damage to the production process and introduce new products.

Figure 20. Innovations in Food Packaging for a Sustainable and Circular Economy

References

He, X.; Deng, H.; Hwang, H.-M. The current application of nanotechnology in food and agriculture. J. Food Drug Anal. **2019**, 27, 1–21. [Google Scholar] [CrossRef]

Singh, R.; Dutt, S.; Sharma, P.; Sundramoorthy, A.K.; Dubey, A.; Singh, A.; Arya, S. Future of Nanotechnology in Food Industry: Challenges in Processing, Packaging, and Food Safety. Glob. Chall. **2023**, 7, 2200209. [Google Scholar] [CrossRef]

Ganeson, K.; Mouriya, G.K.; Bhubalan, K.; Razifah, M.R.; Jasmine, R.; Sowmiya, S.; Amirul, A.-A.A.; Vigneswari, S.; Ramakrishna, S. Smart packaging—A pragmatic solution to approach sustainable food waste management. Food Packag. Shelf Life **2023**, 36, 101044. [Google Scholar] [CrossRef]

Adeyemi, J.O.; Fawole, O.A. Metal-Based Nanoparticles in Food Packaging and Coating Technologies: A Review. Biomolecules **2023**, 13, 1092. [Google Scholar] [CrossRef]

Sharma, A.; Ranjit, R.; Pratibha; Kumar, N.; Kumar, M.; Giri, B.S. Nanoparticles based nanosensors: Principles and their applications in active packaging for food quality and safety detection. Biochem. Eng. J. **2023**, 193, 108861. [Google Scholar] [CrossRef]

Aswathi, V.P.; Meera, S.; Maria, C.G.A.; Nidhin, M. Green synthesis of nanoparticles from biodegradable waste extracts and their applications: A critical review. Nanotechnol. Environ. Eng. **2022**, 8, 377–397. [Google Scholar] [CrossRef]

Hamad, A.F.; Han, J.-H.; Kim, B.-C.; Rather, I.A. The intertwine of nanotechnology with the food industry. Saudi J. Biol. Sci. **2018**, 25, 27–30. [Google Scholar] [CrossRef]

Alfadul, S.M.; Elneshwy, A.A. Use of nanotechnology in food processing, packaging and safety—Review. Afr. J. Food Agric. Nutr. Dev. **2010**, 10, 6. [Google Scholar] [CrossRef]

Mitrano, D.M.; Wohlleben, W. Microplastic regulation should be more precise to incentivize both innovation and environmental safety. Nat. Commun. **2020**, 11, 5324. [Google Scholar] [CrossRef]

Bumbudsanpharoke, N.; Ko, S. Nano-Food Packaging: An Overview of Market, Migration Research, and Safety Regulations. J. Food Sci. **2015**, 80, R910–R923. [Google Scholar] [CrossRef]

Amenta, V.; Aschberger, K.; Arena, M.; Bouwmeester, H.; Moniz, F.B.; Brandhoff, M.P.; Gottardo, S.; Marvin, H.J.P.; Mech, A.; Pesudo, L.Q.; et al. Regulatory aspects of nanotechnology in the agri/feed/food sector in EU and non-EU countries. Regul. Toxicol. Pharmacol. **2015**, 73, 463–476. [Google Scholar] [CrossRef]

Bodbodak, S.; Rafiee, Z. Recent trends in active packaging in fruits and vegetables. In Eco-Friendly Technology for Postharvest Produce Quality; Elsevier: Amsterdam, The Netherlands, 2016; pp. 77–125. [Google Scholar]

Thite, N.G.; Ghazvini, S.; Wallace, N.; Feldman, N.; Calderon, C.P.; Randolph, T.W. Machine Learning Analysis Provides Insight into Mechanisms of Protein Particle Formation Inside Containers During Mechanical Agitation. J. Pharm. Sci. **2022**, 111, 2730–2744. [Google Scholar] [CrossRef]

Yam, K.L.; Takhistov, P.T.; Miltz, J. Intelligent packaging: Concepts and applications. J. Food Sci. **2005**, 70, R1–R10. [Google Scholar] [CrossRef]

Stoma, M.; Dudziak, A. Eastern Poland Consumer Awareness of Innovative Active and Intelligent Packaging in the Food Industry: Exploratory Studies. Sustainability **2022**, 14, 13691. [Google Scholar] [CrossRef]

Fadiji, A.E.; Mthiyane, D.M.N.; Onwudiwe, D.C.; Babalola, O.O. Harnessing the Known and Unknown Impact of Nanotechnology on Enhancing Food Security and Reducing Postharvest Losses: Constraints and Future Prospects. Agronomy **2022**, 12, 1657. [Google Scholar] [CrossRef]

Kuswandi, B.; Jumina. Active and intelligent packaging, safety, and quality controls. In Fresh-Cut Fruits and Vegetables: Technologies and Mechanisms for Safety Control; Academic Press: Cambridge, MA, USA, 2020; pp. 243–294. [Google Scholar]

Fang, Z.; Zhao, Y.; Warner, R.D.; Johnson, S.K. Active and intelligent packaging in meat industry. Trends Food Sci. Technol. **2017**, 61, 60–71. [Google Scholar] [CrossRef]

Ghaani, M.; Cozzolino, C.A.; Castelli, G.; Farris, S. An overview of the intelligent packaging technologies in the food sector. Trends Food Sci. Technol. **2016**, 51, 1–11. [Google Scholar] [CrossRef] [Green Version]

Han, J.H.; Ho, C.H.L.; Rodrigues, E.T. Intelligent packaging. In Innovations in Food Packaging; Elsevier: Amsterdam, The Netherlands, 2005; pp. 138–155. [Google Scholar]

Rodrigues, C.; Souza, V.G.L.; Coelhoso, I.; Fernando, A.L. Bio-based sensors for smart food packaging—Current applications and future trends. Sensors **2021**, 21, 2148. [Google Scholar] [CrossRef]

Wyrwa, J.; Barska, A. Innovations in the food packaging market: Active packaging. Eur. Food Res. Technol. **2017**, 243, 1681–1692. [Google Scholar] [CrossRef]

Alves, J.; Gaspar, P.D.; Lima, T.M.; Silva, P.D. What is the role of active packaging in the future of food sustainability? A systematic review. J. Sci. Food Agric. **2023**, 103, 1004–1020. [Google Scholar] [CrossRef]

Roy, S.; Rhim, J.-W. Carboxymethyl cellulose-based antioxidant and antimicrobial active packaging film incorporated with curcumin and zinc oxide. Int. J. Biol. Macromol. **2020**, 148, 666–676. [Google Scholar] [CrossRef]

Chen, S.; Brahma, S.; Mackay, J.; Cao, C.; Aliakbarian, B. The role of smart packaging system in food supply chain. J. Food Sci. **2020**, 85, 517–525. [Google Scholar] [CrossRef] [Green Version]

Rodrigues, C.; de Paula, C.D.; Lahbouki, S.; Meddich, A.; Outzourhit, A.; Rashad, M.; Pari, L.; Coelhoso, I.; Fernando, A.L.; Souza, V.G.L. Opuntia spp.: An Overview of the Bioactive Profile and Food Applications of This Versatile Crop Adapted to Arid Lands. Foods **2023**, 12, 1465. [Google Scholar] [CrossRef]

Perez Espitia, P.J.; de Fátima Ferreira Soares, N.; dos Reis Coimbra, J.S.; de Andrade, N.J.; Souza Cruz, R.; Alves Medeiros, E.A. Bioactive Peptides: Synthesis,

Properties, and Applications in the Packaging and Preservation of Food. Compr. Rev. Food Sci. Food Saf. **2012**, 11, 187–204. [Google Scholar] [CrossRef]

Nanda, S.; Berruti, F. Thermochemical conversion of plastic waste to fuels: A review. Environ. Chem. Lett. **2021**, 19, 123–148. [Google Scholar] [CrossRef]

Siracusa, V.; Blanco, I. Bio-Polyethylene (Bio-PE), Bio-Polypropylene (Bio-PP) and Bio-Poly(ethylene terephthalate) (Bio-PET): Recent Developments in Bio-Based Polymers Analogous to Petroleum-Derived Ones for Packaging and Engineering Applications. Polymers **2020**, 12, 1641. [Google Scholar] [CrossRef]

Mangaraj, S.; Yadav, A.; Bal, L.M.; Dash, S.K.; Mahanti, N.K. Application of Biodegradable Polymers in Food Packaging Industry: A Comprehensive Review. J. Packag. Technol. Res. **2019**, 3, 77–96. [Google Scholar] [CrossRef]

Alshehrei, F. Biodegradation of Synthetic and Natural Plastic by Microorganisms. J. Appl. Environ. Microbiol. **2017**, 5, 8–19. [Google Scholar]

Vieira, I.R.S.; de de Carvalho, A.P.A.; Conte-Junior, C.A. Recent advances in biobased and biodegradable polymer nanocomposites, nanoparticles, and natural antioxidants for antibacterial and antioxidant food packaging applications. Compr. Rev. Food Sci. Food Saf. **2022**, 21, 3673–3716. [Google Scholar] [CrossRef]

Scarfato, P.; Di Maio, L.; Incarnato, L. Recent advances and migration issues in biodegradable polymers from renewable sources for food packaging. Appl. Polym. Sci. **2015**, 132, 42597. [Google Scholar] [CrossRef]

Salgado, P.R.; Di Giorgio, L.; Musso, Y.S.; Mauri, A.N. Recent Developments in Smart Food Packaging Focused on Biobased and Biodegradable Polymers. Front. Sustain. Food Syst. **2021**, 5, 630393. [Google Scholar] [CrossRef]

Haider, T.P.; Völker, C.; Kramm, J.; Landfester, K.; Wurm, F.R. Plastics of the Future? The Impact of Biodegradable Polymers on the Environment and on Society. Angew. Chem. Int. Ed. **2019**, 58, 50–62. [Google Scholar] [CrossRef] [Green Version]

Chausali, N.; Saxena, J.; Prasad, R. Recent trends in nanotechnology applications of bio-based packaging. J. Agric. Food Res. **2022**, 7, 100257. [Google Scholar] [CrossRef]

Peelman, N.; Ragaert, P.; De Meulenaer, B.; Adons, D.; Peeters, R.; Cardon, L.; Van Impe, F.; Devlieghere, F. Application of bioplastics for food packaging. Trends Food Sci. Technol. **2013**, 32, 128–141. [Google Scholar] [CrossRef] [Green Version]

King, T.; Cole, M.; Farber, J.M.; Eisenbrand, G.; Zabaras, D.; Fox, E.M.; Hill, J.P. Food safety for food security: Relationship between global megatrends and developments in food safety. Trends Food Sci. Technol. **2017**, 68, 160–175. [Google Scholar] [CrossRef]

Nazari, M.; Majdi, H.; Gholizadeh, P.; Kafil, H.S.; Hamishehkar, H.; Zarchi, A.A.K.; Khoddami, A. An eco-friendly chitosan/cellulose acetate hybrid nanostructure containing Ziziphora clinopodioides essential oils for active food packaging applications. Int. J. Biol. Macromol. **2023**, 235, 123885. [Google Scholar] [CrossRef] [PubMed]

Alizadeh-Sani, M.; Mohammadian, E.; McClements, D.J. Eco-friendly active packaging consisting of nanostructured biopolymer matrix reinforced with TiO_2 and essential oil: Application for preservation of refrigerated meat. Food Chem. **2020**, 322, 126782. [Google Scholar] [CrossRef] [PubMed]

Venezia, V.; Prieto, C.; Evtoski, Z.; Marcoaldi, C.; Silvestri, B.; Vitiello, G.; Luciani, G.; Lagaron, J.M. Electrospun hybrid TiO_2/humic substance PHBV films for active food packaging applications. J. Ind. Eng. Chem. **2023**, 124, 510–522. [Google Scholar] [CrossRef]

Fabra, M.J.; Lopez-Rubio, A.; Lagaron, J.M. High barrier polyhydroxyalcanoate food packaging film by means of nanostructured electrospun interlayers of zein. Food Hydrocoll. **2013**, 32, 106–114. [Google Scholar] [CrossRef]

FSC
www.fsc.org
MIX
Papier | Fördert gute Waldnutzung
FSC® C083411

Kolibri 360 GmbH
Ferdinand-Jühlke-Straße 7
99095 Erfurt, Deutschland
produktsicherheit@kolibri360.de

Druck:
CPI Druckdienstleistungen GmbH
im Auftrag der
Kolibri 360 GmbH
Ferdinand-Jühlke-Str. 7
99095 Erfurt